Caroline Styles

Caroline Styles

Magnetic Resonance Imaging and Computed Tomography of the Head and Neck

We dedicate this book to our wives and children,

Denise, Jenny, Mark and David; and
Joanne, Bethan and Nicholas

for their support and tolerance during the preparation of this project.

J. E. Gillespie
A. Gholkar

Magnetic Resonance Imaging and Computed Tomography of the Head and Neck

Edited by
James E. Gillespie
Consultant Neuroradiologist, Manchester Royal Infirmary, and
Lecturer in Diagnostic Radiology, University of Manchester,
Manchester, UK

and

Anil Gholkar
Consultant Neuroradiologist, Newcastle General Hospital,
and hon. lecturer in Diagnostic Radiology, University of Newcastle
upon Tyne, Newcastle upon Tyne, UK

CHAPMAN & HALL MEDICAL
London · Glasgow · Weinheim · New York · Tokyo · Melbourne · Madras

Published by Chapman & Hall, 2–6 Boundary Row, London SE1 8HN, UK

Chapman & Hall, 2–6 Boundary Row, London SE1 8HN, UK

Blackie Academic & Professional, Wester Cleddens Road, Bishopbriggs, Glasgow G64 2NZ, UK

Chapman & Hall Inc., One Penn Plaza, 41st Floor, New York NY10119, USA

Chapman & Hall Japan, Thomson Publishing Japan, Hirakawacho Nemoto Building, 6F, 1-7-11 Hirakawa-cho, Chiyoda-ku, Tokyo 102, Japan

Chapman & Hall Australia, Thomas Nelson Australia, 102 Dodds Street, South Melbourne, Victoria 3205, Australia

Chapman & Hall India, R. Seshadri, 32 Second Main Road, CIT East, Madras 600 035, India

First edition 1994

© 1994 Chapman & Hall

Typeset in 10/12 Palatino by Keyset Composition, Colchester, Essex
Printed and bound in Hong Kong

ISBN 0 412 45200 6

A catalogue record for this book is available from the British Library

Library of Congress Cataloging-in-Publication data available

Contents

Contributors

Bernadette M. Carrington
Lecturer in Oncological Radiology
Manchester University
Christie Hospital
Manchester

Richard A. Fawcitt MB ChB, FRCR
Department of Neuroradiology
Manchester Royal Infirmary
Oxford Road
Manchester M13 9WL, UK

W. St Claire Forbes FRCR
Consultant Neuroradiologist
Hope Hospital, Eccles Old Road
Salford M6 8HD, UK

Anil Gholkar MB BS, FRCR
Consultant Radiologist
Department of Neuroradiology
Regional Neurosciences Centre
Newcastle General Hospital
Newcastle upon Tyne, UK

James E. Gillespie MB BCh, BAO, DMRD, FRCR
Consultant Radiologist
Department of Neuroradiology
Manchester Royal Infirmary
Oxford Road
Manchester M13 9WL, UK

Stephen J. Golding
Clinical Director
Oxford MRI Centre
Oxford, UK

Ian Holland
Consultant Neuroradiologist
Queen's Medical Centre
Nottingham NG7 2UH, UK

Alan Jackson BSc, PhD, MB ChB, MRCP, FRCR
Department of Neuroradiology
Manchester Royal Infirmary
Oxford Road
Manchester M13 9WL, UK

Tim Jaspan
Consultant Neuroradiologist
Queen's Medical Centre
Nottingham NG7 2UH, UK

Richard J. Johnson
Director of Diagnostic Radiology
Christie Hospital
Manchester

Foreword

In years past head and neck disorders have sometimes been perceived as occupying a no-man's land between the territories of the Neurosurgical and the ENT Departments, with anatomical and pathological diagnoses often achieved only by surgical dissection. Radiology has been a remarkable catalyst in the development of modern multi-disciplinary approaches to the head and neck. It has also come to occupy a central and often decisive role in the clinical management of diseases affecting this vital area.

Perhaps the most significant paradigm shift in scientific radiology during the 20th century occurred as a consequence of the introduction of computed tomography in 1972. Computed tomography had a profound effect on diagnostic medicine, introducing the concepts of digital data acquisition, interactive displays and powerful image processing to *in vivo* biological activities. A wide range of previously inaccessible clinical situations became available to the radiologist for diagnosis and for treatment planning, nowhere more so than in the head and neck. Computed tomography stimulated the scientific environment for the major developments to follow. Magnetic resonance proton imaging, with its high contrast multi-planar facilities and absence of ionizing radiation, is currently the most sensitive and least invasive modality for the investigation of structural change particularly of the soft tissues. It is now established as a primary investigative tool in a wide range of clinical conditions and is generally accepted as a crucial diagnostic facility in the investigation of head and neck disorders. The exquisite delineation of structure by both computed tomography and magnetic resonance imaging has permitted logical *in vivo* analyses of morphology, function and pathology, and is providing in turn a deeper and richer understanding of the nature and spread of disease.

It is important that all those whose efforts are directed towards disorders affecting the head and neck should understand the role of modern imaging, and the opportunities that it provides to influence the management and enhance the welfare of the afflicted patient. This is a timely volume, and for those in training especially it offers both a broad perspective and a detailed account of the imaging techniques and their application.

It gives me great pleasure and pride to be invited to write this Foreword. Both distinguished editors trained in Manchester, and both held Research Fellowships in the University Department to and from which they contributed much important work.

I. Isherwood

Preface

Standard textbooks of radiology cover the radiographic aspects of head and neck investigation well but, usually due to lack of space, can give only limited coverage of CT and MRI. In practice, the anatomical complexity and range of pathology in this area frequently necessitates the use of cross-sectional imaging, often as the primary imaging investigation. At the other end of the spectrum, there are several excellent reference texts on head and neck imaging which fulfil the needs of those with a specialist interest in this field but which are too detailed for the non-specialist. We felt there was a need to fill this gap with a relatively short but well-illustrated book dealing with the application of CT and MRI in the investigation of head and neck disease.

In the first chapter general imaging principles are discussed from a practical viewpoint. Thereafter, the anatomy and pathology of the head and neck from the skull base superiorly to the larynx inferiorly, as shown by CT and MRI, are presented in six chapters. We have included as much of the clinical aspects of disease as was feasible in a book of this size. The role of plain radiography and other methods of investigation are discussed briefly in the context of overall radiological investigation but are not described in detail nor illustrated as this information is available elsewhere. Intracranial lesions above the skull base, and cervical spine pathology, are not included.

While this book is primarily written for radiologists in training and general radiologists undertaking occasional CT or MRI studies in this area, we hope clinicians specializing in skull base, ENT, maxillo-facial and plastic surgery will also find it valuable.

J. E. Gillespie
A. Gholkar

1 Imaging principles

Anil Gholkar and James E. Gillespie

INTRODUCTION

Imaging of head and neck disease is a challenging task. This region is divided into several subdivisions. However, many of the anatomical structures and pathological processes involve more than one of these divisions. The role of imaging is to make a diagnosis as well as to define the exact extent of the lesion so that appropriate surgical or radiotherapy treatment can be given. Mucosal changes are best evaluated by physical examination, but the deep extent of the lesion can only be demonstrated by imaging.

Imaging modalities used to assess head and neck disease are:

1. plain radiographs
2. conventional tomography
3. ultrasound
4. computed tomography
5. magnetic resonance imaging.

We will mainly discuss CT and MRI in this chapter.

Plain radiographs

These are often performed as the initial investigation in suspected inflammatory or neoplastic paranasal sinus disease and in patients with facial trauma. This may be the only investigation needed in certain clinical situations such as acute sinusitis or isolated nasal bone fractures (Chapter 2). However, their role in imaging of head and neck disease is limited. For a detailed description of different views the reader is referred to the recommended reading at the end of this chapter.

Conventional tomography

Prior to the availability of computed tomography, conventional tomography was used to assess the extent of bony abnormality and soft tissue asymmetry. However, nowadays, conventional tomography does not add any information to that provided by computed tomography and should not be performed.

Ultrasound

The use of ultrasound is mainly limited to imaging of orbits. In the assessment of ocular and some retro-ocular lesions, ultrasound can be extremely useful (Chapter 4).

COMPUTED TOMOGRAPHY (CT)

Computed tomography has been used for assessing head and neck disease for several years. A detailed description of CT physics can be obtained in some of the larger CT textbooks. The following discussion will deal with the application of CT scanning techniques to the head and neck.

CT scanning techniques

In order to obtain the maximum amount of information possible there are five aspects of the CT examination which must be carefully considered prior to the commencement of the examination.

Image planes required

Many abnormalities in the head and neck need to be viewed in more than one plane for optimum assessment. This can be accomplished either by:

1. single plane (normally axial) scanning with subsequent multiplanar reformations (MPR); or
2. scanning directly in multiple planes (direct multiplanar scanning).

Each of these techniques will now be considered in turn.

Single plane scanning

With MPR a series of continuous CT slices are obtained with a slice width as narrow as possible. 1.5 mm sections or less are ideal; 3 mm sections are acceptable in many situations but widths greater than 3 mm will not produce satisfactory reformation.

Advantages of MPR:

1. The patient position for axial scanning is comfortable and stable.
2. The patient is scanned only once and so is irradiated only once.
3. The single set of scan data obtained can be subsequently reformated into any two-dimensional plane or be used to produce three-dimensional reformations (see below).

Disadvantages of MPR:

1. The patient must be co-operative enough to maintain a constant position throughout the entire examination to avoid misregistration artefact.

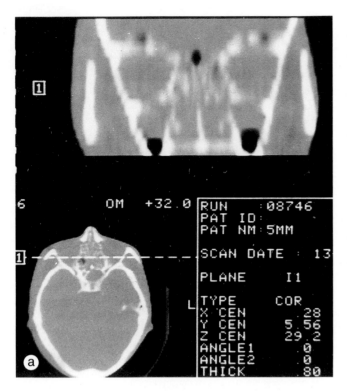

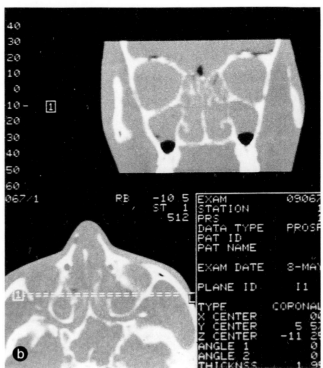

Fig. 1.1 CT reformation: effect of slice width on spatial resolution. (a) Coronal reformation from 5 mm thick slices; (b) coronal reformation from 1.5 mm thick slices. Both images are at mid-orbital level in a skull phantom. Bony detail is vastly superior in (b).

2. Spatial resolution of reformated images is normally inferior to that of a directly obtained CT section. Spatial resolution is related to image pixel size and the pixels in a CT slice are square. Pixels in a reformated image are rectangular in shape, the length of the rectangle being dependent upon the slice width of the original CT slices. Therefore, reducing the slice width of the axial slices to be reformated will improve the spatial resolution of any subsequent reformations. This effect of differing slice widths is illustrated in Fig. 1.1. As well as reducing slice width, further improvement in spatial resolution can be obtained by overlapping the axial sections. Radiation dose to the patient, however, is increased and the authors prefer contiguous sections. While accepting the technically inferior spatial resolution of MPR in comparison to directly obtained scans, this does not necessarily mean that MPR will produce clinically inferior images (Fig. 1.2).
3. Modern CT scan consoles will produce reformations rapidly but the operator must nevertheless spend at least some time initiating the desired images.

Direct multiplanar scanning

Direct multiplanar scanning involves manoeuvring the patient's head into different positions with respect to the scanner gantry. The image planes possible by this method depend mainly upon the flexibility of the patient's neck and his/her ability to maintain this position for the required length of time.

Advantages of direct multiplanar scanning:

1. Optimal spatial resolution.
2. Patient movement between individual scans is tolerable providing movement does not occur during data acquisition.
3. No reformation time is required after the examination.

Disadvantages of direct multiplanar scanning:

1. Difficult patient positioning. In practice, direct coronal (or near coronal) scans can be achieved in most patients with a reasonable range of neck movement. Tilting the gantry can help achieve the desired plane. However, direct sagittal or sagittal oblique scans are difficult or impossible to perform in many patients. Consequently the image planes achievable are more limited than with MPR.
2. Performing direct scans in two or more planes will significantly increase the radiation dose to

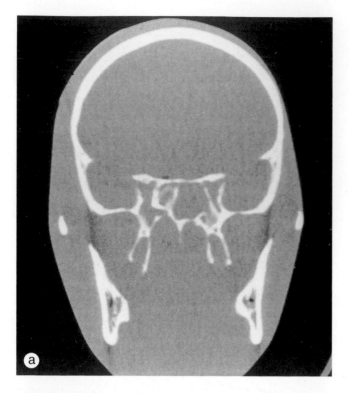

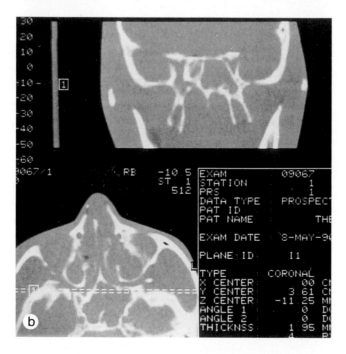

Fig. 1.2 Direct multiplanar scanning versus multiplanar reformations. (**a**) 1.5 mm direct coronal scan; (**b**) coronal reformation from axial 1.5 mm slices. Both images are from a skull phantom at approximately the same anatomical level. The bony detail of the pterygoid plates and sphenoid sinus region in (**b**) is comparable to that in (**a**) and satisfactory for clinical purposes.

the patient. This is a serious disadvantage of the direct multiplanar technique particularly when scanning near radiation-sensitive organs such as the eye and thyroid gland.

3. Streak artefact from dental fillings can significantly degrade CT images and is a common problem with attempted direct coronal scans. Adjustments in gantry tilt or head positioning to avoid the fillings may circumvent the problem but the 'coronal' image achieved will often be coronal-oblique.

Slice width selection

Several factors must be considered in selecting the most appropriate slice width for the clinical problem being investigated:

1. **Total body area that needs to be examined**. If a large anatomical area must be covered it may be necessary to use thicker slices than would otherwise be desired due to constraints of examination time and heat generation within the X-ray tube.

2. **Size of the lesion or structure to be visualized**. In general terms, slice width should approximate the size of the structure to be visualized. If the object of interest is much smaller than the CT slice employed, definition will be reduced or lost due to the partial volume effect.

3. **Use of MPR**. If reformated images are required then appropriate slice widths must be used (see above).

4. **Scan technique to be employed**. If the dynamic scan mode is to be used (see below) then the number of sections that can be acquired at a given mAs will be limited by the amount of heat generated within the X-ray tube and its cooling rate. Complete coverage of a given area in dynamic scan mode may necessitate either using thicker sections than originally intended or reducing the mAs level so that a larger number of narrower slices can be obtained without incurring tube cooling delays.

Contrast enhancement

Intravenous contrast injection has two major purposes in the head and neck:

1. To highlight areas of pathology.
2. To allow differentiation of vascular from non-vascular structures. This is essential in order to

avoid mistaking soft tissue masses such as lymph nodes with vessels seen in cross-section. Information regarding vascular patency and invasion or displacement by adjoining masses will also be provided.

A simple bolus of contrast prior to scanning as is commonly used in the investigation of cerebral mass lesions is usually inadequate as vessel conspicuity falls away rapidly during the course of the examination. Enhancement must be obtained before the scan commences and plasma contrast levels maintained during the examination. There are many variations of this technique but most involve giving an initial bolus of contrast followed by contrast infusion. One suitable technique is to infuse 100 ml of contrast at a rate of $2\,\mathrm{ml\,s^{-1}}$ using an injection pump, and to commence scanning 15 s after the start of injection. The authors favour the use of non-ionic contrast as it is better tolerated by the patient than ionic media, at a concentration of 300 mg iodine per millilitre.

As it is highly desirable to scan the entire area of interest while enhancement is at a maximum, the infusion technique is best combined with dynamic (incremental) scanning (Fig. 1.3).

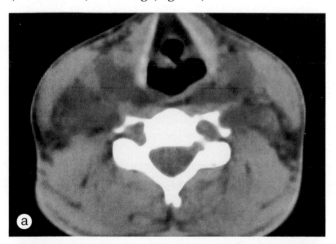

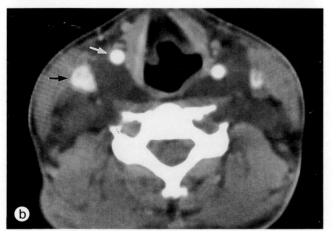

Normal or rapid acquisition (dynamic) scan mode

Most CT scanners have a rapid data acquisition scan mode, most commonly called dynamic scanning. Dynamic scan mode differs from normal scan mode in that the X-ray tube anode does not stop rotating between slices and so scanning is interrupted only by the time required for table movement to a new location (dynamic incremental scanning) or by whatever pre-selected time delay between scans at the same location is desired (dynamic sequential scanning). Reconstruction of the image data is either delayed until all scans have been performed, or else proceeds concurrently with scanning, although the images are not viewed until the end of the examination.

Dynamic sequential scanning can provide detailed, quantifiable information regarding rates of enhancement over time but has generally little application in most clinical circumstances. Dynamic incremental scanning, however, is an extremely useful technique, allowing the greatest number of scans to be performed in the least time possible. This can very usefully be combined with intravenous contrast infusion, as described above.

An even more rapid means of scanning a large volume of patient anatomy, spiral CT scanning, has recently become available. Its value in a variety of clinical areas is currently under investigation, including its use in combination with intravenous contrast injection and 3D reformating (spiral CT angio graphy) [1].

Fig. 1.3 Dynamic contrast enhancement. Axial CT scans at hypopharyngeal level in a patient with neurofibromatosis. On the plain scan (**a**) multiple well defined neurofibromata are identified, hypodense to surrounding muscle. The position of the major neck vessels is unclear in (**a**) but in (**b**), obtained dynamically during intravenous contrast infusion, the common carotid arteries (white arrow) and internal jugular veins (black arrow) are clearly visualized. Note the separation of the vein and artery by the neurofibroma within the carotid sheath bilaterally.

Scanning factors/radiation dose

For a scan obtained at a given voltage and at a pre-set slice width and reconstruction matrix, the signal-to-noise ratio and the radiation dose to the patient will vary proportionately to changes in the mAs employed. Increasing the mAs used for a scan will thus not only improve image quality but also increase the radiation dose to the patient as well as increasing the amount of heat generated within the X-ray tube. For example, studies performed using a phantom on a GE9800 CT scanner [2] demonstrated that the radiation dose at 140 mAs was approximately half that produced at 300 mAs (Table 1.1). Therefore, by reducing the mAs to the lowest level consistent with obtaining diagnostic images, the radiation dose to the patient can be minimized. For many soft tissue pathologies in the head and neck the authors commonly perform scans at 140 mAs with very satisfactory clinical results (see Fig. 6.41, p.178). Furthermore, using reduced mAs levels will allow a greater number of scans to be obtained in dynamic mode, unimpeded by X-ray tube heating constraints.

Table 1.1   Radiation dose to the skin surface in 3D CT (reproduced from reference 2, with permission)

Examination	mAs/kVp	Radiation dose (mGy) ±6%
Craniofacial	300	40
	140	19
	80	11
Pelvic	300	19
	80	5

The principles discussed above are applicable generally throughout the head and neck. Specific applications of these principles to individual pathologies are discussed in the forthcoming chapters.

Three-dimensional (3D) reformating of CT data

The MPR technique described previously produces two-dimensional images and this facility is widely available on most CT scanners as standard. Advances in computer software now enable digital data to be displayed three-dimensionally using a variety of software methods. Three-dimensional reformating can be performed either on the CT scan computer itself, usually as an optional software addition, or as part of a dedicated image-processing workstation. The latter is the more expensive option but usually allows 3D reformating of both bone and soft tissue structures and a greater degree of image manipulation and interaction. Such systems can handle digital data not only from CT but also from magnetic resonance and isotope studies.

The 3D software which can run on the CT scanner's own computer is usually less expensive but has more limited capabilities. Reformating is mainly limited to high-density bony structures and has found its main applications in the investigation of congenital and traumatic abnormalities of the skull vault, facial skeleton and spine. Soft tissue 3D reformating is rather limited and largely restricted to enhancing intracranial structures. Three-dimensional images produced by these CT-based systems can be rotated through 360° around any axis on the display monitor in real or near-real time. Viewer appreciation of the three-dimensional effect is accomplished by two main methods. The first, called 'depth-encoding' uses the CT grey-scale to denote distance rather than density and so points closer to the observer appear brighter than those which are further away. In the second main method, called 'surface-reflectance', brightness is changed according to the inclination of the object surface, giving a greater degree of surface detail. These two methods of display are shown in Fig. 1.4. In practice a combination of both methods is usually employed. For a detailed discussion of the technical and clinical aspects of 3D imaging the reader is referred to the text by Udupa and Herman.

Three-dimensional reformations provide an excellent means of conveying complex spatial information to the observer in an accurate, easily recognized format. Extracting three-dimensional relationships from a series of two-dimensional images in areas of anatomical complexity can be testing, especially for relatively inexperienced viewers. Three-dimensional images thus complement the original two-dimensional CT images from which they are derived. Examples of this complementary relationship are shown in Chapter 2 where mainly depth-encoded 3D images are used to display complex facial fractures.

The scan requirements for good quality 3D reformations are identical to those described for multiplanar reformations. Narrow slice widths and a motionless patient are essential. Dynamic incremental scanning at reduced mAs levels is ideally suited to 3D reformating [1]. Spiral CT data acquisition may also be employed.

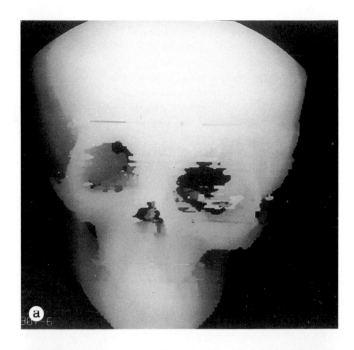

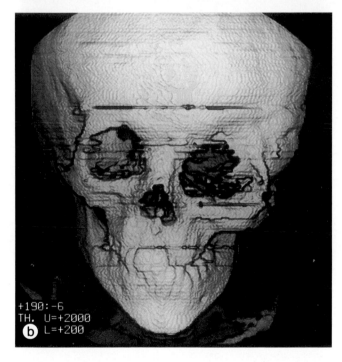

Fig. 1.4 3D surface reformations. Anterior 3D reformation of a child with a left-sided malar fracture displayed in pure (**a**) depth-encoded and (**b**) surface-reflectance modes. Irregularities of the orbital outlines and forehead are due to slice misregistration artefact by slight changes in head position between scans. (Reproduced by courtesy of Booth Hall Childrens Hospital, Manchester.)

MAGNETIC RESONANCE IMAGING (MRI)

Since it became available, MRI has replaced CT as the investigation of choice in many head and neck lesions. The advantages of MRI over CT are given in Table 1.2. The most significant advantages are lack of ionizing radiation, multiplanar imaging and superb soft tissue contrast. The main disadvantages of MRI are limited availability and cost. Small numbers of patients cannot complete the examination due to claustophobia. Patients with metallic foreign bodies, prostheses and cardiac pacemakers are contraindicated in MRI.

Table 1.2 Advantages of MRI

High intrinsic soft tissue contrast
Direct multiplanar imaging
No ionizing radiation
No bone and air artefacts
No biological hazards

Physical principles

The physics of MRI are discussed in detail elsewhere [3,4]. The MR signal is produced by atomic nuclei with an odd number of proton and/or neutrons (hydrogen(^{1}H), sodium (^{23}Na), phosphorus (^{31}P) etc.). These nuclei act as bar magnets. When placed in a static magnetic field, application of a radiofrequency pulse can induce resonance in particular sets of nuclei. Release of energy occurs as the RF pulse is turned off is detected by a receiving coil and converted to an electric signal which provides data for a digital image.

Magnet

At the heart of the MRI system is a magnet providing a stable and homogeneous static magnetic field. Field strengths for different clinical MRI systems range from 0.02–2.0 tesla. The advantage of higher-field-strength systems is the greater amount of signal produced. This 'extra' signal can be translated into faster scan times (increased patient throughput), thinner slices (increased spatial resolution) or simply better quality (less 'noisy') images than would be possible on a lower-field-strength unit. High-field systems, however, are more costly to purchase and more prone to various image artefacts (e.g. chemical shift, magnetic susceptibility).

Radiofrequency coils

These coils surround the parts of the patient to be examined in order to transmit and receive radiofrequency signal to and from the patient. A head coil is frequently used to image the orbits and paranasal sinuses. However, smaller-diameter surface coils, which can be closely applied to the patient, improve the signal-to-noise ratio.

Surface coils also allow thinner sections to be obtained from a smaller field of view with resultant improvement in spatial resolution and soft tissue contrast.

MR tissue parameters

Unlike CT, which has only one parameter (attenuation coefficient), MR has at least ten tissue parameters which may contribute to an image (Table 1.3). The two most frequently used tissue parameters are the T1 and T2 relaxation times. T1-weighted images show excellent anatomical detail while T2-weighted images are more sensitive in detection of pathology.

Table 1.3 MRI tissue parameters

Proton density
T1 relaxation time
T2 relaxation time
Chemical shift
Flow effect
Susceptibility
Diffusion
Perfusion
Radiofrequency absorption

Pulse sequences

Pulse sequences allow you to manipulate the contribution to image contrast made by different tissue parameters. Partial saturation, spin echo and inversion recovery are commonly used pulse sequences [4]. Gradient echo sequences are sometimes substituted for spin echo sequences due to their shorter imaging time (Fig. 7.4, p.199). They have advantages in detection of calcium and haemorrhage, but are more prone to susceptibility artefacts. Within a sequence, a number of machine-based variables (e.g. TR, TE, flip angle, etc) can be altered to change the contrast between different tissue components.

Most lesions have a high signal intensity on T2-weighted images and low/intermediate signal intensity on T1-weighted images (Fig. 1.5). Normal fat in the subcutaneous tissue or orbits is of high signal intensity on T1-weighted images. This contrasts well with pathological lesions which are of low to intermediate signal intensity on unenhanced T1-weighted images. However, on T2-weighted images high fat signal may reduce the contrast between fat and pathology. Fat suppression is a useful technique to demonstrate lesions in certain parts of head and neck and can be accomplished using 1) short tau inversion recovery (STIR) or 2) chemical shift imaging.

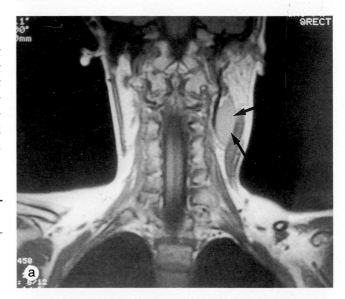

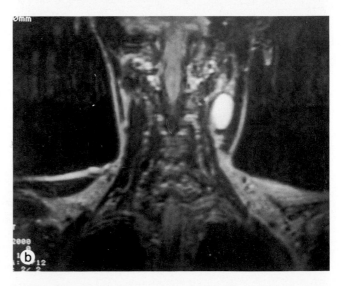

Fig. 1.5 Coronal images of the neck in a patient with branchial cyst. (**a**) The lesion is of intermediate signal intensity on T1-weighted image (arrow) and of high signal intensity on T2-weighted image (**b**).

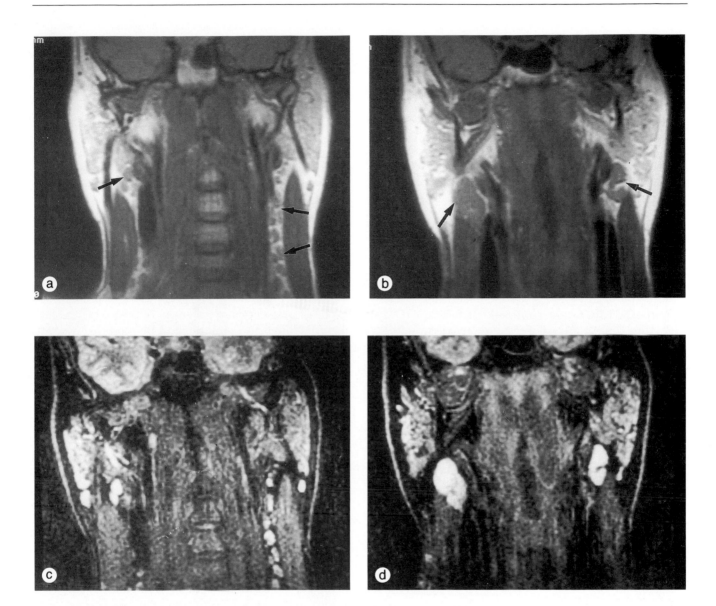

Fig. 1.6 A case of metastatic cervical lymphadenopathy. T1-weighted coronal images (**a**, **b**) show multiple bilateral nodes (arrowed). The nodes are much more conspicuous on STIR images (**c**, **d**), due to high signal intensity of lesion and suppression of signal from surrounding fat.

Short tau inversion recovery (STIR)

This provides T1 and T2 contrast together with cancellation of the signal from fat. Pathological lesions are therefore of high signal intensity, contrasted against low signal intensity from surrounding fat and muscle.

This makes it a very useful sequence in imaging orbits. Pathological cervical lymph nodes are also well demonstrated by this sequence (Fig. 1.6). However it is important to note that paramagnetic contrast injection results in decreased signal intensity of STIR images due to negative contrast enhancement.

Chemical shift imaging

Chemical shift effects occur due to a change in the MR behaviour of protons relating to differences in the chemical environment between protons attached to carbon in a lipid molecule and those bound to oxygen, as in water.

This difference can be exploited in chemical shift imaging to produce separate water and lipid or composite (phase contrast) images. Several different methods (Dixon method, chopper fat suppression sequence, Sepponen method, etc.) use this technique to achieve fat suppression. These sequences demonstrate positive contrast enhancement after injection of paramagnetic contrast (Fig. 4.5, p. 67).

Magnetic resonance angiography

MR images are very sensitive to flow phenomenon. Flowing blood can either appear bright or dark depending on velocity [5]. The other factors influencing signal from blood flow are the pulse sequence and MR imager. MR angiography is a relatively recent development which uses flow quantitation techniques to isolate the vascular system from the surrounding stationary tissues. Extracranial carotid vessels can be demonstrated well using MR angiography.

Contrast agents

Paramagnetic contrast agents are used intravenously to enhance MR tissue contrast. Gadolinium DTPA was the first contrast medium used in clinical practice. More recently other gadolinium-based compounds have become available for use. These paramagnetic agents affect the local tissue magnetic susceptibility. They shorten the T1 relaxation time, producing increased signal on T1-weighted images.

Contrast enhancement occurs in certain normal tissues such as the nasal mucosa and the pituitary gland.

Contrast agents will enhance inflammatory and neoplastic lesions, but the most useful information from enhancement is obtained when there is extension into the intracranial compartment. There are certain disadvantages of contrast enhancement as the lesions may become difficult to separate from surrounding fat or normally enhancing structures such as pituitary gland.

Protocols

Some of the common protocols used in imaging of the head and neck are given in Table 1.4. These are just guidelines and may need to be modified according to the requirements of each examination.

Table 1.4 Protocols for MRI of the head and neck (SE = spin echo; STIR = short tau inversion recovery; T1W = T1-weighted; T2W = T2-weighted; c = paramagnetic contrast)

Region	Pulse sequence	Imaging plane
Skull base	SE T1W	Axial ± c
	SE T1W	Coronal ± c
	SE T2W	Axial
Orbits	SE T1W	Axial
	SE T1W	Coronal ± c
	STIR	Coronal
Paranasal sinuses	SE T1W	Axial
	SE T1W	Coronal ± c
	SE T2W	Axial
Nasopharynx	SE T1W	Axial ± c
	SE T1W	Sagittal
	SE T1W	Coronal ± c
	STIR	Coronal
Oropharynx	SE T1W	Axial
	SE T1W	Coronal ± c
	STIR	Coronal
Hypopharynx	SE T1W	Axial
	SE T1W	Sagittal
	SE T1W	Coronal
	STIR	Coronal

REFERENCES

1. Schwartz RB, Jones KM, Chernoff DM et al. Common carotid artery bifurcation: evaluation with spiral CT. Radiology 1992; 185: 513–519.
2. Gholkar A, Gillespie JE, Hart CW, Mott D, Isherwood I. Dynamic low-dose three dimensional computed tomography: a preliminary study. British Journal of Radiology 1988; 61: 1095–1099.
3. Smith MA. The technology of Magnetic Resonance Imaging. Clinical Radiology 1985; 36: 553–559.
4. Bradley WG, Bydder GM, Worthington BS. Magnetic resonance imaging: basic principles. In: Grainger RG, Allison DJ, eds. Diagnostic radiology, 2nd ed. Edinburgh: Churchill Livingstone, 1992.
5. Mills CM, Brant-Zawadzki M, Crooks LE et al. Nuclear magnetic resonance: principles of blood flow imaging. American Journal of Neuroradiology 1983; 4: 1161–1166.

RECOMMENDED READING

Stark DD, Bradley WA. Magnetic resonance imaging. St Louis, MO: CV Mosby/Year Book, 1992

Udapa JK, Herman G. 3D imaging in medicine. Boca Raton, FL: CRC Press, 1991.

Valvassori GE, Potter GD, Hanafee WN, Carter BL, Buckingham RA (eds). Radiology of the ear, nose and throat. Philadelphia, PA: WB Saunders, 1984.

2 Facial trauma

James E. Gillespie

INTRODUCTION

The facial bones are one of the most important and complex portions of skeletal anatomy. They provide the structural support and points of attachment for the facial soft tissues, divide the face into distinct compartments and protect vital soft tissue organs from injury. Craniofacial injuries may result in cosmetic disfigurement and adversely affect basic biological functions such as vision and nutrition. Prompt assessment and treatment are essential.

The majority of facial fractures result from assault, road traffic accidents and sports injuries. Since the introduction of seat belt legislation and stricter drink/driving laws [1], the incidence of facial injuries from road accidents has been declining. Sport injuries, on the other hand, are increasing and are now the commonest cause of serious eye injury in the UK [2].

RADIOLOGICAL INVESTIGATION

The purpose of radiological investigation is to provide the surgeon with enough information to plan optimal patient treatment. Most patients with suspected facial bone injury will first undergo conventional radiography to confirm the presence of a fracture. This may be all the imaging required for isolated nasal bone or mandibular fractures and for associated dental injuries. Superimposition of bony structures and impaired visualization of underlying fractures by soft tissue swelling and haemorrhage, however, may necessitate further imaging in more complicated injuries. Linear or complex-motion tomography has been used to help define sites of fracture in such circumstances. However, tomography provides none of the soft tissue information now considered essential in planning treatment, especially around the orbit, and carries a significant radiation penalty to the patient. With the increasing use of open reduction and internal fixation techniques, details of the degree of comminution and displacement of bony fragments is essential in guiding surgical exposure and fixation [3]. High-resolution computed tomography (CT), with its excellent soft tissue and bone definition, is undoubtedly the imaging modality of choice for the detailed evaluation of facial injury. Magnetic resonance imaging (MRI) has, at present, only a minor role in the investigation of facial trauma, which will be discussed later.

APPLIED ANATOMY OF THE FACIAL SKELETON

Traditionally, the face has been divided into an upper third comprising the frontal bone, a lower third comprising the mandible and a middle third extending from the supraorbital margins to the maxillary alveolus. Gentry et al. [4, 5] conceptualized the upper two-thirds of the face as three interconnected groups of bony struts orientated in the sagittal, coronal and horizontal planes. Each group contains two or three subdivisions based upon location: the sagittal group contains median, para-sagittal and lateral struts, the coronal group contains anterior and posterior struts, and the horizontal group has superior, middle and inferior struts (Fig. 2.1). Tables 2.1–2.3 detail the major osseous strut components, along with important adjacent soft tissue structures and potential complications that may ensue following strut fracture.

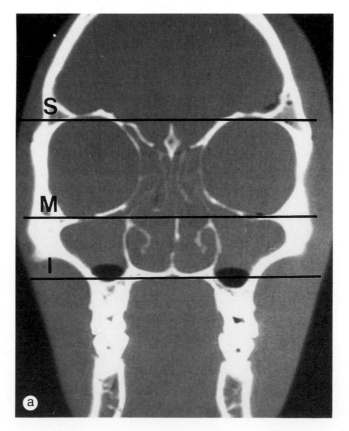

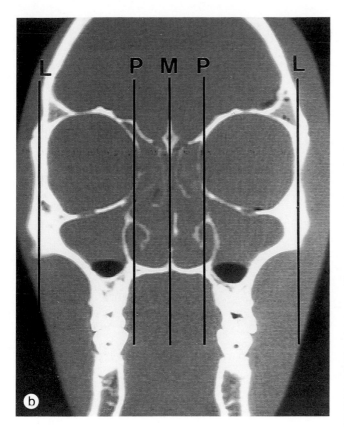

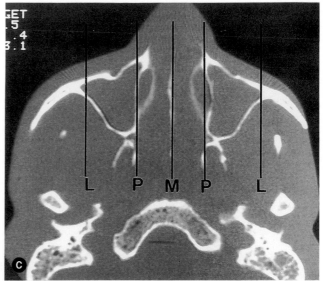

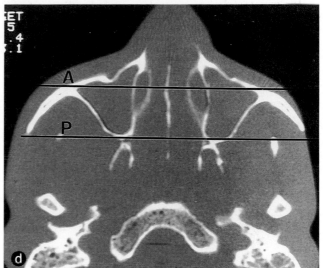

Fig. 2.1 **Facial strut positions on CT**. (**a**) Coronal scan showing the superior (S), middle (M) and inferior (I) horizontal struts. (**b**) and (**c**) Coronal and axial scans respectively showing the median (M), parasagittal (P) and lateral (L) sagittal struts. (**d**) Axial scan demonstrating the anterior (A) and posterior (P) coronal struts.

Table 2.1   Horizontal plane struts (adapted from reference 5)

Osseous components	Associated soft tissues	Potential complications
1. SUPERIOR Orbital roof and cribriform plate	Brain, meninges, olfactory and optic nerves, upper extraocular muscles, anterior and posterior ethmoidal arteries and nerves	Intracranial injury, anosmia, blindness, CSF leak (infection), muscle entrapment, epistaxis
2. MIDDLE Orbital floor	Lower extraocular muscles, orbital fat, infraorbital artery and nerve, eye	Muscle entrapment, fat prolapse, infraorbital nerve damage (sensory impairment), ocular injury
Zygomatic arch	Temporalis muscle	Muscle injury
3. INFERIOR Hard palate	Incisive and palatine foramen, arteries	Haematoma, oronasal fistulae

Table 2.2 **Coronal plane struts** (adapted from reference 5)

Osseous components	Associated soft tissues	Potential complications
1. ANTERIOR		
Frontal	Brain, meninges, supraorbital artery and nerve	Intracranial injury and infection, sinusitis, mucocele, telecanthus, diplopia, epiphora, chronic dacrocystitis
Nasofrontal	Lacrimal sac, nasofrontal duct, medial canthal ligament	
Zygomaticofrontal	Lacrimal gland, lateral canthal ligament	
Anterior alveolar	(Anterior) superior dental artery and nerve	Devitalized teeth
2. POSTERIOR		
Posterior maxillary sinus wall Pterygoid plates	Sphenopalatine, pterygopalatine and internal maxillary arteries, trigeminal nerve (maxillary division), sphenopalatine ganglion	Retropharyngeal haematoma (respiratory compromise), internal maxillary artery thrombosis or pseudoaneurysm, maxillary nerve injury, chronic facial pain, eustachian tube blockage (middle ear infection)

Table 2.3 **Sagittal plane struts** (adapted from reference 5)

Osseous components	Associated soft tissues	Potential complications
1. MEDIAN		
Nasal septum	Nasopalatine artery and nerve	Nasal septal haematoma (leading to septal deviation and perforation)
2. PARASAGITTAL		
Medial orbital wall	Medial rectus and superior oblique muscles, trochlea of superior oblique, anterior and posterior ethmoidal artery and nerve	Enophthalmos and diplopia, superior oblique tendon sheath syndrome (extraocular muscle dysfunction), periorbital haematoma and epistaxis
Medial maxillary sinus wall	Nasolacrimal duct, maxillary sinus ostium	Chronic sinusitis, retention mucocele
3. LATERAL		
Lateral orbital wall	Superior and inferior orbital fissure contents, lateral rectus muscle, lacrimal gland, middle cranial fossa contents	Superior orbital fissure syndrome (damage to cranial nerves III, IV, VI and V by bone or haematoma), infraorbital numbness, lateral rectus injury, intracranial injury
Lateral maxillary sinus wall Lateral alveolar ridge	(Posterior) superior dental artery and nerve	Devitalized teeth

Figures 2.2 and 2.3 depict the normal bony anatomy of the face as demonstrated by CT using a commercially produced skull phantom. Note that axial images demonstrate the sagittal and coronal struts well but show the horizontal struts relatively poorly; coronal images demonstrate the horizontal and sagittal struts to advantage but the coronal struts less well. While it must be remembered that many facial injuries will involve more than one strut, use of this geometric conceptualization provides one means of evaluating bony injury systematically and anticipating soft tissue complications.

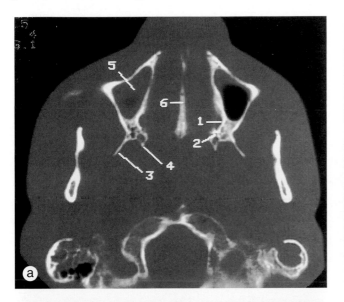

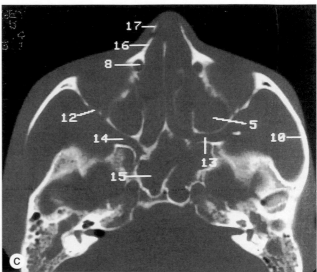

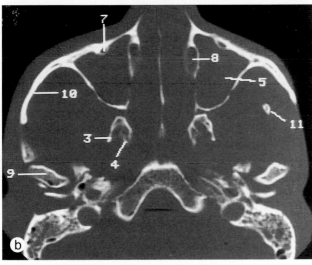

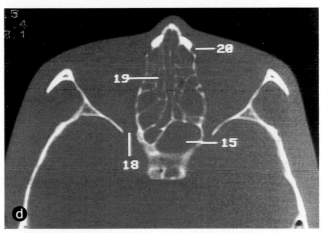

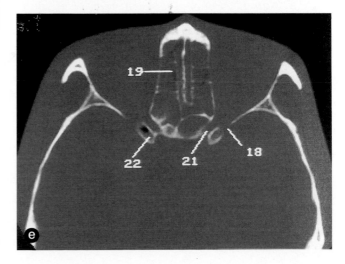

Fig. 2.2 Normal axial osseous anatomy. Axial 1.5 mm CT sections (skull phantom) at the level of the lower (**a**), mid (**b**) and upper (**c**) maxillary antra, superior orbital fissure (**d**) and optic foramina (**e**).

1 = greater palatine foramen; 2 = lesser palatine foramen;
3 = lateral pterygoid plate; 4 = medial pterygoid plate;
5 = maxillary sinus; 6 = vomer; 7 = infraorbital canal;
8 = nasolacrimal duct; 9 = mandibular condyle;
10 = zygomatic arch; 11 = coronoid process; 12 = inferior orbital fissure; 13 = pterygopalatine fossa;
14 = pterygomaxillary fissure; 15 = sphenoid sinus;
16 = frontal process of maxilla; 17 = nasal bone;
18 = superior orbital fissure; 19 = ethmoid sinus;
20 = lacrimal fossa; 21 = optic canal; 22 = anterior clinoid.

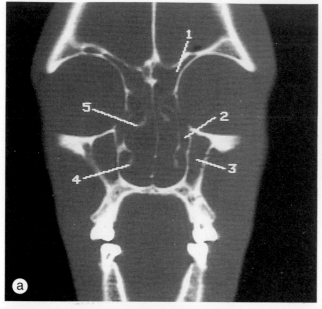

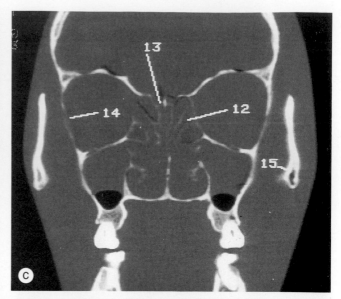

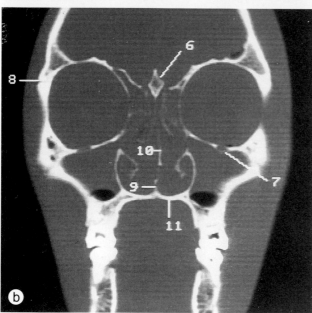

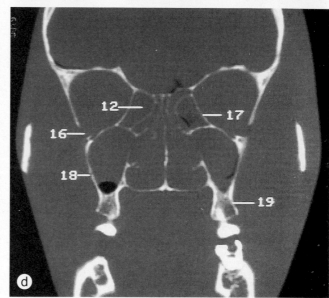

Fig. 2.3 Normal coronal osseous anatomy. Coronal
1.5 mm CT sections (skull phantom) progressing from the
anterior (**a**) to the posterior (**e**) coronal strut levels. (**b**), (**c**)
and (**d**) are within the anterior, middle and posterior thirds
of the orbit respectively.

1 = frontal sinus; 2 = nasolacrimal duct; 3 = maxillary
sinus; 4 = inferior turbinate; 5 = middle turbinate;
6 = crista galli; 7 = infraorbital canal; 8 = frontozygomatic
suture; 9 = vomer; 10 = perpendicular plate of ethmoid;
11 = hard palate; 12 = ethmoid sinus; 13 = cribriform
plate; 14 = spenozygomatic suture; 15 = zygoma;
16 = inferior orbital fissure; 17 = orbital plate of the
ethmoid; 18 = alveolar canal; 19 = alveolar process of
maxilla; 20 = anterior clinoid; 21 = optic canal;
22 = superior orbital fissure; 23 = foramen rotundum;
24 = sphenoid sinus; 25 = lateral pterygoid plate;
26 = medial pterygoid plate; 27 = zygomatic arch.

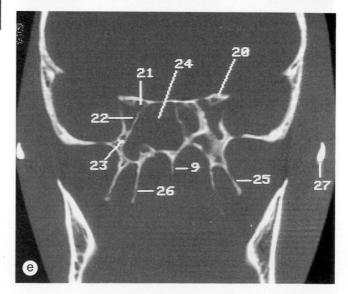

CT SCANNING TECHNIQUE

As indicated previously, any one image plane will demonstrate two groups of struts well, but not usually the third. Given the possible consequences of orbital and anterior cranial fossa trauma, assessment of horizontal and sagittal strut integrity is of paramount importance. Therefore, any imaging protocol must at least produce coronal images, as this is the best single plane to show both these strut groups [4]. Posterior displacement cannot be assessed adequately on coronal images, however, and consequently images in two or more planes are needed for an optimal three-dimensional appraisal of facial injury.

The choice between thin-slice axial imaging with subsequent image reformating and direct multiplane scanning will depend upon individual clinical circumstances and the CT facilities available. Both techniques have advantages and disadvantages (Chapter 1). When modern fast CT facilities are available with good reformating capabilities, the author's preference is to use the axial slice/reformating technique, as it has significant advantages over direct multiplane scanning in most circumstances. Patient positioning for axial scans has the advantage of both comfort and stability and may be the only choice possible in acute trauma; scan time and radiation dose are kept to a minimum, as only one set of images need to be acquired; and examination of the brain or cervical spine, when required, can proceed without position change. This is important as up to 33% of patients with severe facial injuries may have significant intracranial injury.

In the author's experience the minor loss of spatial resolution incurred by reformating is not clinically significant, even in the assessment of orbital soft tissue complications. This view is supported by other authors [6, 7], though direct multiplane scanning has its advocates [8]. Apart from the facility to derive conventional multiplanar images from a single set of scan data, three-dimensional (3D) reformations can be generated if suitable software is available. These can be a valuable adjunct in moderate and severe facial injuries by providing a clearer perception of the extent of major fracture lines and, in particular, the degree of bony displacement [9].

Axial scans are normally obtained parallel to the anthropological baseline (external auditory meatus to inferior orbital margin). Optimum quality reformations require as narrow a slice width as possible. Ideally, this should be 1.5 mm or less, particularly around the orbits (Figs 2.4–2.8, 2.14, 2.16) but should not exceed 3 mm to maintain acceptable spatial resolution both in the axial scans and reformations (Fig. 2.20).

Slice widths can be combined, e.g. contiguous 1.5 mm sections through the orbits with 3 mm slices through the maxilla or frontal bones (Fig. 2.10). Scans should be obtained using a dynamic low-dose technique (Chapter 1) to reduce examination time and radiation dose [10]. For example, 60 slices can be obtained at 80 mAs in just over 5 minutes on most modern scanners. This protocol will produce scans with excellent osseous and adequate soft tissue detail for facial injury assessment. Direct coronal scans are aligned perpendicular to the hard palate, although in practice the image plane obtained is often coronal-oblique due to constraints imposed by dental fillings and the degree of neck extension the patient can comfortably maintain. A 3 mm slice width is usual.

CLASSIFICATION OF FRACTURES

Facial fractures can be classified into two basic groups [11]:

1. **Transfacial fractures** comprising the Le Fort I, II and III level fractures, in which a portion of the mid-face becomes detached from the skull base, is freely mobile and must be stabilized surgically;
2. **Limited fractures**, in which bony injury is more localized and no major detachment of the facial bones from the skull base occurs.

Limited fractures can be sub-divided into:

a) **Single strut injuries** involving one of the orbital walls zygomatic arch, paranasal sinus wall or nasal bones; and
b) **Multiple strut injuries** of the zygomatic complex ('tripod' fractures), nasofrontal and nasoethmoidal regions.

Mandibular fractures are usually classified separately and will be discussed later.

Fracture classification serves to convey to clinicians a general description of the site of major fracture lines and the level of any instability present. However, once a general fracture classification has been made, the radiologist must proceed with a detailed analysis of the bony and soft tissue injuries present as these may vary significantly from patient to patient, even in similarly classified injuries.

LIMITED FRACTURES

Single strut injuries

Isolated fractures of the nasal bones and zygomatic arch can usually be identified on plain radiographs and do not normally require CT. Their appearance on CT, along with sinus wall fractures, will be demonstrated later when they occur in conjunction with other injury patterns. The main single strut fractures requiring CT are those of the orbit.

The site and pattern of fractures around the orbit is influenced by the direction, point of impact and magnitude of the striking force, as well as the bony anatomy of the orbital walls. Superiorly and laterally the orbit has relatively thick bony walls. Inferiorly and medially, however, the orbit is separated from the maxillary antrum and ethmoidal sinus respectively by thin plates of bone which are more vulnerable to injury. In this section, single strut fractures of the orbital walls will be discussed. It should be remembered that significant orbital injury can also occur in fractures of the zygomatic complex, nasofrontal and nasoethmoidal regions, as well as Le Fort II and III level fractures. These will be discussed in the appropriate sections.

Orbital floor

The classic injury of the orbital floor is the 'blow-out' fracture whose postulated mechanism of injury is as follows.

When the anterior part of the globe is struck by a blunt object, usually slightly larger than the orbital inlet, the orbital contents are displaced backwards, thus acutely raising intraorbital pressure; the sudden increase in hydraulic pressure fractures the weakest portions of the orbit, namely the orbital floor and/or the medial wall, with orbital soft tissues being forced through the bony defects. This is the 'pure' blow-out fracture in which the anterior orbital rim is intact. The herniated orbital soft tissues lying within the upper antrum give rise to the classic 'hanging-drop' opacity. However, the floor can also be fractured by a blunt force striking the inferior orbital rim, resulting in a buckling of the bony floor with subsequent fracture, the so-called 'impure' blow-out fracture.

A number of fracture patterns can be seen. Comminution with several small depressed fragments may be noted or a single fragment may be seen hinged to one side of the fractured floor like a trap-door. The majority of fractures occur medial to the infraorbital canal or groove. When the anterior orbital rim is involved, the fracture line extends posteriorly in continuity and may involve a significant length of the floor.

Visualization of the orbital floor and adjacent soft tissues requires images at right angles to the horizontal plane. Coronal images are the minimum required to assess this area. However, sagittal-oblique images parallel to the inferior rectus muscle display the entire muscle in continuity and its relationship to the fractured floor. This valuable view can be obtained easily by reformatting (Fig. 2.4) or with greater difficulty by direct scanning.

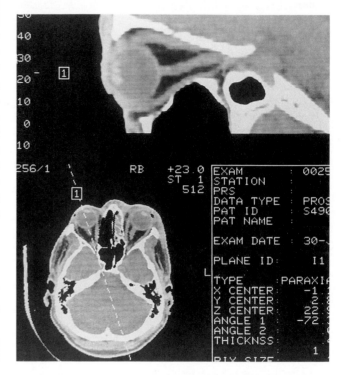

Fig. 2.4 Normal orbit. Sagittal-oblique reformatted CT image. The reformatting plane is along the optic nerve/ inferior rectus muscle and was generated from a series of contiguous 1.5 mm axial slices. Note that the entire length of the inferior rectus muscle and its relationship to the orbital floor are displayed in continuity. The optic nerve and superior rectus muscle are also well visualized.

Complications

The major complications of orbital floor fractures are diplopia and enophthalmos. Direct entrapment or incarceration of the inferior oblique and inferior rectus muscles at the fracture site was long considered to be the main cause of diplopia, but CT has shown this to be uncommon. A system of connective tissue septa exists between the orbital walls, the muscles and the globe [8]. This fine ligament system interconnects all orbital soft tissues and is most developed at eyeball level, becoming less dense posteriorly. When the orbital floor is fractured and fat herniates through the defect the connective tissue system becomes caught within the fracture site and the muscles are indirectly entrapped or tethered. If left untreated, adhesions may form within the connective tissue system, fixing the muscle to the orbital floor.

Diplopia can be present in up to 72% of patients with blow-out fractures. In many patients this is transient, the motility disorder becoming permanent in 25% if left untreated. On the basis of CT imaging, four mechanisms of motility impairment have been defined [8]:

1. post-traumatic oedema/haemorrhage
2. post-traumatic oculomotor nerve palsy
3. gross enophthalmos
4. muscle entrapment/adhesions.

The distinction between these is important as diplopia due to the first two causes will probably resolve with eye movement exercises. When the latter two causes are present, however, surgical treatment is necessary to prevent permanent motility impairment. CT in post-traumatic oedema/haemorrhage demonstrates an area of soft tissue density at the fracture site near the inferior rectus muscle, which may be swollen (Fig. 2.5) or displaced upwards by the abnormal soft tissue (Fig. 2.6) but with no signs of incarceration or tethering.

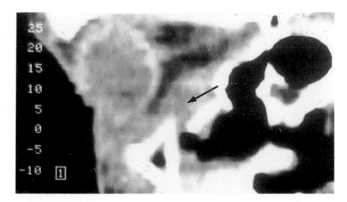

Fig. 2.6 Orbital floor blow-out fracture. Sagittal-oblique CT reformation. The patient had undergone orbital surgery one week previously but had residual diplopia. The scan demonstrates that the inferior rectus muscle is displaced upwards by a heterogeneous soft tissue mass representing haematoma (arrow). Note the depressed orbital floor fragment immediately below.

Post-traumatic oculomotor nerve palsy can be inferred when the scan shows no other significant cause of diplopia and there are other clinical signs of oculomotor involvement, such as anisocoria. In gross enophthalmos, which will be discussed below, the posterior displacement of the globe causes the rectus muscles to become relatively slack and this impairs efficient muscle contraction. In muscle entrapment there is an angular displacement of the muscle belly; incarceration is said to be present when the angulated muscle lies within the fracture line itself (Fig. 2.7), whereas in tethering the angulated muscle lies above the fracture site (Fig. 2.8).

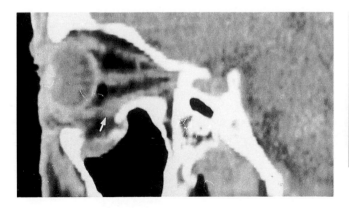

Fig. 2.5 Orbital floor blow-out fracture. The sagittal-oblique CT reformation demonstrates swelling of the inferior rectus muscle near the posterior edge of the floor fracture (arrow). Note the pocket of air lying below the optic nerve/globe junction and herniation of orbital fat, bone and haemorrhagic soft tissue into the upper antrum.

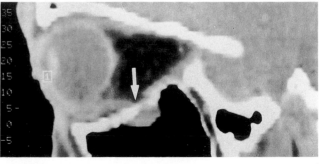

Fig. 2.7 Orbital floor blow-out fracture. Sagittal-oblique reformated CT image. The inferior rectus muscle is incarcerated within the fractured floor (arrow).

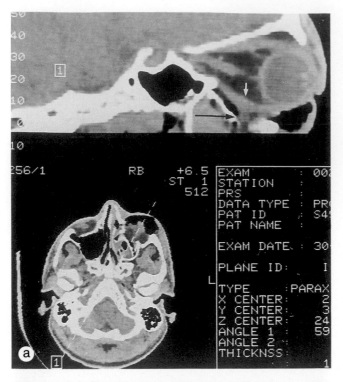

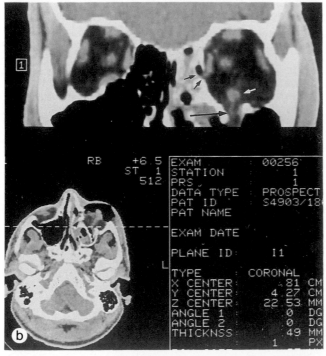

Adhesions are seen as abnormal soft tissue extending between the fracture site and muscle, with loss of definition of muscle margins and obliteration of the normal surrounding fat planes.

Enophthalmos, defined as a backward and usually downward displacement of the globe into the orbit, occurs in up to 14% of patients with orbital fractures. It may present acutely or develop progressively. Globe position is determined by three structures: the bony orbit, the ligament system and orbital fat [12]. The orbit is basically cone-shaped with the medial and inferior walls bulging inward behind the globe. When a significant fracture occurs the space available for the orbital soft tissues enlarges and becomes more spherical in shape while the soft tissue volume itself remains more or less constant. If ligamentous disruption accompanies this increase in bony orbital volume, the soft tissues become displaced and altered in shape allowing enophthalmos to develop. Any subsequent scar tissue formation and contracture may worsen the situation.

Surgical correction is aimed at restoring orbital volume and shape and may include orbital floor prosthesis insertion (Fig. 2.15). Marked orbital enlargement and globe displacement will be obvious clinically as well as on CT. With lesser degrees of enlargement, quantitative assessment of orbital volume may help predict which patients are liable to develop enophthalmos and will therefore require surgery to prevent its development [13].

Medial wall

The incidence of medial wall blow-out fractures has been underestimated in the past as plain radiographs often show some ethmoidal clouding but not usually the fracture itself. Studies using cadaver orbits have shown a high incidence of concomitant medial wall fractures when orbital floor blow-outs have been induced experimentally [14]. Greater use of CT in orbital injuries has confirmed this higher incidence, up to 50% of patients with orbital floor blow-outs also having a similar injury of the medial wall. A medial wall blow-out may also occur on its own. While small defects of the thin lamina papyracea may not be of great clinical importance, larger

Fig. 2.8 Orbital floor and medial wall blow-out fractures. (**a**) Sagittal-oblique reformated CT image. The inferior orbital rim is displaced posteriorly and the orbital floor fractured along its entire length. The inferior rectus muscle (white arrow) has lost its bony support and is sagging downwards. Ill-defined linear soft tissue strands representing early adhesions are seen extending between the muscle belly and fracture site (black arrow). Haemorrhagic fluid is present in the upper antrum and

subcutaneous air is visible near the lower eyelid.
(**b**) Coronal reformation. The increase in left orbital volume is obvious along with the downwardly displaced inferior rectus muscle (small white arrow) and early adhesions extending from its inferior margin (long black arrow). Note the 'trap-door' configuration of the medial side of the floor fracture. A small medial wall blow-out is also evident (small black arrows) near the inferior margin of the medial rectus muscle.

fractures can be associated with significant soft tissue prolapse into the ethmoidal sinus and cause or contribute to the development of enophthalmos and ocular movement disorders. CT images in the axial or coronal plane will readily demonstrate these fractures and any associated soft tissue complications (Fig. 2.9).

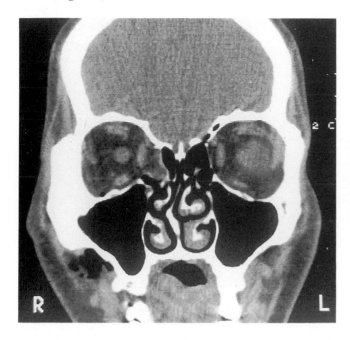

Fig. 2.9   Medial wall blow-out fracture. Direct coronal CT scan showing a large medial wall blow-out fracture of the right orbit with fat prolapse into the ethmoid sinus. No displacement of the medial rectus muscle is seen.

Orbital roof

The orbital roof is a thin structure but is reinforced laterally by the greater wing of the sphenoid and anteriorly by the superior orbital rim. It is therefore much more resistant to sudden increases in intraorbital pressure than the floor or medial wall. Blow-out fractures have been described in patients with large frontal sinuses extending posteriorly over the orbital roof, but these are unusual. Rarely, a sudden increase in intracranial pressure from a distant skull fracture may decompress through the orbital roof, this being the so-called 'blow-in' fracture (Fig. 2.28). Most orbital roof fractures, however, are associated with severe craniofacial trauma, with blunt force being applied to the region around the superior orbital margin, causing the roof to buckle and fracture. The fracture often involves the weaker medial portion of the roof with extension into the frontal bone and cribriform plate (Fig. 2.10).

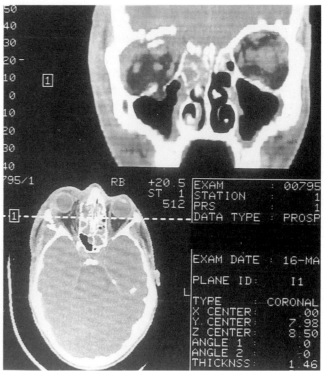

Fig. 2.10   Orbital roof fracture. Coronal CT reformation demonstrating a comminuted right orbital roof fracture. Herniation of brain through the defect is present. The patient has also sustained orbital floor, cribriform plate and zygomatic complex fractures.

Lateral wall

This is the strongest wall of the orbit, being formed by the dense zygomatic bone and greater wing of the sphenoid. Fractures of the lateral wall usually occur as part of a zygomatic complex (tripod) fracture and will be described in the next section.

Timing of CT in blow-out fractures

There is general agreement that, if a blow-out fracture requires surgical treatment, it should be performed sooner rather than later as delayed surgical treatment of adhesions and enophthalmos is difficult and associated with a less satisfactory outcome. The indications for, and the timing of, CT in a patient with a blow-out fracture vary from place to place. In some centres CT is reserved for those with an absolute indication for surgery, i.e., diplopia failing to resolve on conservative treatment, or the development of enophthalmos. Other centres perform CT early in a majority of patients and use the information to help select those cases likely to require surgery.

Multiple strut injuries

Zygomatic complex (tripod fracture)

The zygoma, or cheekbone, forms much of the lateral wall and inferior rim of the orbit and articulates with the frontal, maxillary, temporal and sphenoid bones. Direct impact on the zygomatic prominence (malar eminence) can result in a series of fractures around the zygoma, potentially isolating it from the adjacent portion of the face. In its complete form, the fracture line involves the frontozygomatic suture and extends downwards and backwards along the lateral orbital wall close to or between the frontal process of the zygoma and the greater wing of the sphenoid to reach the anterior aspect of the inferior orbital fissure; the fracture line then extends anteromedially to the inferior orbital margin near the infraorbital canal, infero-laterally across the antrum beneath the zygomatic prominence, then upwards across the posterior antral wall to the inferior orbital fissure anteriorly (Fig. 2.11). Separation of the zygo-ma from the face is completed by a fracture through the zygomatic arch at its weakest point, posterior to the junction between the temporal process of the zygoma and the zygomatic process of the temporal bone (Fig. 2.12).

After impact, the zygoma is usually displaced posteriorly and, depending upon the direction and point of impact, rotated around the vertical and/or horizontal axes (Fig. 2.13). Occasionally, the typical

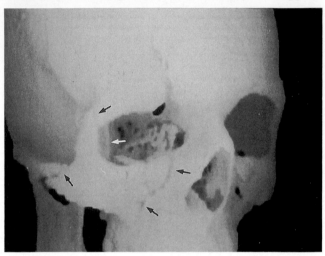

Fig. 2.12 Zygomatic complex fracture. Antero-lateral 3D CT image. The complete fracture pattern is present with the zygomatic arch, antral and lateral orbital components arrowed. Additional frontal bone and orbital roof injuries were sustained. (Source: Gillespie JE, Quayle AA, Barker G, Isherwood I. British Journal of Oral and Maxillofacial Surgery 1987; 25: 175, with permission.)

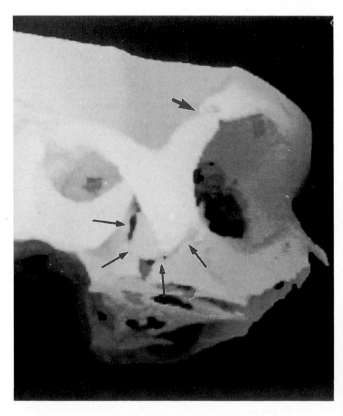

Fig. 2.11 Zygomatic complex fracture. Infero-lateral 3D CT image. The antral component of the fracture, extending from the inferior orbital margin, around the zygomatic prominence and up the posterior antral wall, is clearly depicted (long black arrows). Separation of the frontozygomatic suture is also evident (short black arrow).

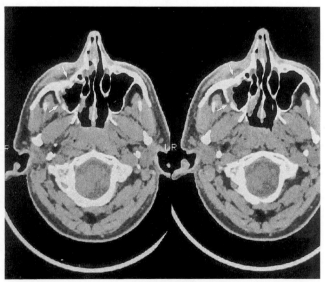

Fig. 2.13 Zygomatic complex fracture. Axial CT sections demonstrating posterior displacement of the right malar eminence with slight clockwise rotation. Fractures of the anterior and lateral antral walls are arrowed.

fracture pattern will be incomplete with the zygomatic arch remaining intact or with incomplete separation at the frontozygomatic suture. The amount of bony displacement in these cases will be less than in the complete form.

CT images in the axial and coronal plane are required to fully assess the degree of malar displacement and comminution, and the extent of orbital injury. Ocular motility impairment and enophthalmos may complicate zygomatic fractures (Fig. 2.14); bony reduction and orbital floor augmentation may both be required to restore normal ocular function and facial appearance (Fig. 2.15). Downward dis-

placement of the lateral canthal ligament which attaches to the frontal process of the zygoma may also occur and cause ocular disfigurement.

Although zygomatic complex fractures commonly occur as isolated injuries, they are also seen in combination with other facial injuries including Le Fort fractures.

Nasofrontal and nasoethmoidal fractures

Both these fracture types are the result of blunt trauma to the central portion of the upper middle third of the face and can be unilateral or bilateral. Nasofrontal fractures involve the supero-medial orbital walls, the frontal sinus and cribiform plate (anterior coronal and superior horizontal struts). Injuries to this site can be associated with significant complications, mainly related to abnormal communication between the intracranial and extracranial compartments and associated orbital injuries (Tables 2.1 and 2.2).

Nasoethmoidal fractures involve the lower two-thirds of the medial orbital walls, ethmoidal sinuses, nasal bones and frontal process of the maxillae (anterior coronal, parasagittal and median sagittal struts). Tables 2.2 and 2.3 list the potential complications, including traumatic telecanthus, which is characteristic. Both fracture types may co-exist (Fig. 2.16).

Assessment of these injuries requires images in the coronal and axial planes. When the site of a dural tear is being sought in patients with CSF rhinorrhoea, a CT cisternogram may be required. Non-ionic water-soluble contrast is injected into the subarachnoid space, usually by lumbar puncture,

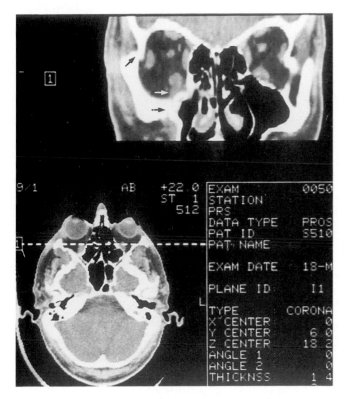

Fig. 2.14 Zygomatic complex fractures: orbital complications. Coronal CT reformation. Fractures of the right lateral and inferior orbital walls are present (black arrows) with downward and outward displacement of the malar fragment with resulting orbital enlargement. Note also the localized haematoma between the fractured floor and the inferior rectus muscle (white arrow).

Fig. 2.15 Zygomatic complex fracture: post-operative repair. Direct coronal CT scan. Bony reduction is incomplete with residual orbital enlargement on the left. A high density Silastic sheet has been inserted to support the orbital soft tissues and prevent the development of enophthalmos.

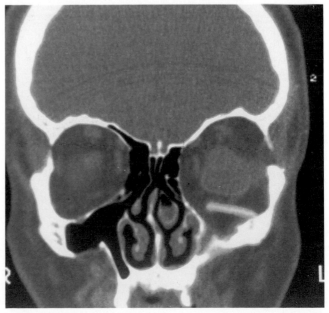

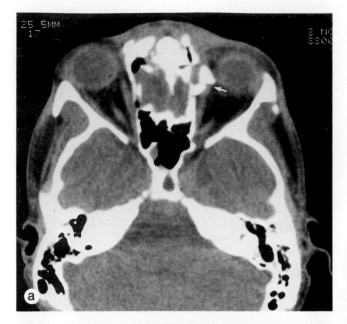

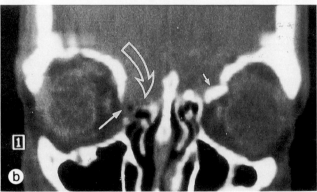

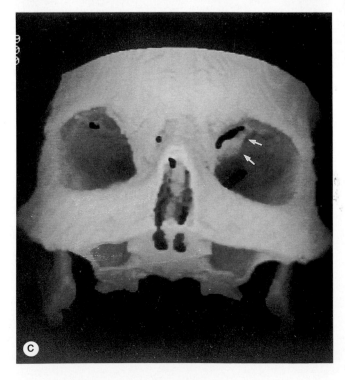

the patient is tilted head down so that contrast enters the head, and direct coronal CT sections are obtained with the patient prone. When a dural tear is present, contrast will enter the part of the paranasal sinuses below the traumatic defect (Fig. 2.17).

TRANSFACIAL FRACTURES

In 1901 René Le Fort described a series of lines along which the facial bones tended to fracture when subjected to significant trauma. Based on this work, transfacial fractures have been divided into three main types:

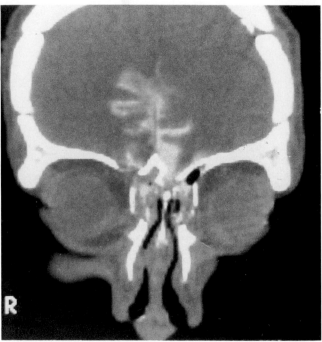

Fig. 2.17   Cribriform plate fractures – CT cisternogram. Direct coronal CT scan after subarachnoid injection of water-soluble non-ionic contrast. High-density contrast is seen entering both ethmoid sinuses through the cribriform plate factures.

Fig. 2.16   Nasofrontal/nasoethmoidal fractures. (**a**) Axial CT scan. Comminuted fractures of the nasofrontal region are present, with bony fragments seen impinging on the left medial rectus muscle (small white arrow). (**b**) Coronal reformation. Bilateral cribriform plate fractures are evident, particularly on the right (curved open arrow), with downward herniation of brain tissue through the defect. A right medial orbital wall fracture is present (long white arrow) as well as the left supero-medial bony fragment shown in (**a**) (small white arrow). (**c**) Antero-inferior 3D CT image. The superior orbital defect and displaced bone fragment are well depicted (arrowed) along with the comminuted nasofrontal and nasoethmoidal fracture lines.

1. **Le Fort I (transverse) fracture** – anteriorly, the fracture line commences below the nasal aperture, extends transversely backwards above the teeth and below the zygoma, across the antero-lateral antral wall, across the pterygomaxillary fossa to fracture the pterygoid plates, and extends to involve the medial antral walls. Thus, the entire lower maxilla and part of the pterygoid plates are detached from the remainder of the face (Fig. 2.18).

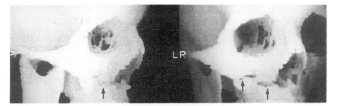

Fig. 2.18 Le Fort I fracture. Lateral-oblique 3D CT images showing the course of the Le Fort I fracture line running transversely across the maxilla (arrows).

2. **Le Fort II fracture** – this is sometimes referred to as a pyramidal fracture because of the shape of the detached facial segment. The apex of the 'pyramid' is at the lower nasal bones; from here the fracture line descends across the frontal process of the maxilla to the inferior orbital margin at the zygomaticomaxillary suture near the infraorbital foramen, extends postero-laterally across the antrum below the malar eminence, then posteriorly to the pterygomaxillary fossa and pterygoid plates. The maxilla, hard palate and pterygoid plates, along with part of the nasoethmoidal bony complex, thus become disassociated from the remainder of the face and skull base (Figs 2.19 and 2.20). In practice, this is the commonest variety of Le Fort fracture.

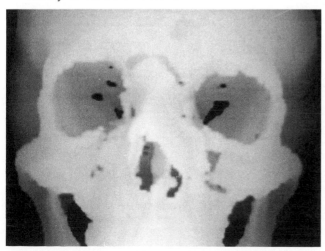

Fig. 2.19 Le Fort II fracture. Anterior 3D CT image showing a classical pyramid-shaped Le Fort II fracture.

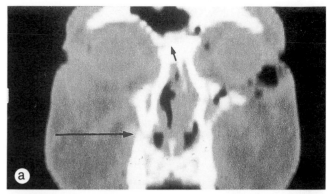

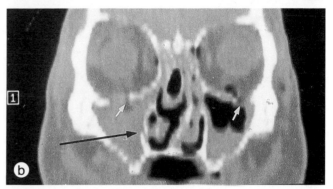

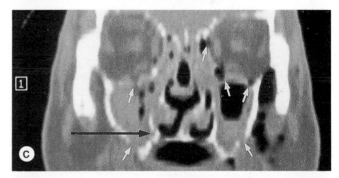

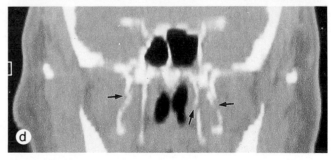

Fig. 2.20 Le Fort II fracture. Series of coronal CT reformations demonstrating the main fracture sites in a Le Fort II fracture. Images extend between the anterior (**a**) and posterior (**d**) coronal struts. Breaches of the nasal bones (**a**), medial and inferior orbital walls, infero-lateral antral margins (**a–c**) and pterygoid plates (**d**) are seen (short arrows). Extensive associated opacification of the ethmoid and maxillary sinuses is present. Note also the right lower medial antral wall and lower nasal septum fractures (**a–c**) indicating a hemi-Le-Fort-I fracture (long arrows).

3. **Le Fort III fracture** – this is the most severe type of transfacial injury, resulting in complete craniofacial disassociation. Commencing at the frontonasal suture, the fracture line extends downwards and backwards, crossing the frontal process of the maxilla, lacrimal bone and medial orbital wall (lamina papyracea of the ethmoids), down to the inferior orbital fissure posteriorly

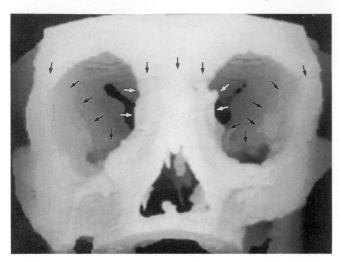

Fig. 2.21   Three-level Le Fort fractures. Anterior 3D CT image of a patient with previously sustained Le Fort I, II and III level fractures. Gross facial distortion is evident despite partial facial reconstruction. Arrows are used to indicate the classical pathway of the Le Fort III fracture line visible in this orientation. In this patient the actual fracture line is better seen in the left orbit than in the right.

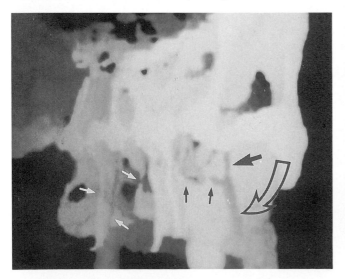

Fig. 2.22   Le Fort II–III level fractures: posterior components. Postero-lateral 3D CT image. The Le Fort II line is seen extending backwards underneath the malar prominence (open curved arrow) and is joined by the Le Fort III line descending from the inferior orbital fissure (large solid arrow) before crossing the pterygoid plates and posterior end of the nasal septum (small arrows).

(Fig. 2.21). From this point the fracture path splits into two components: one line continues upwards across the lateral orbital wall along the sphenozygomatic and frontozygomatic sutures (as in zygomatic complex fractures); the other line descends across the posterior antral wall near the pterygomaxillary fossa and fractures the pterygoid plates near the sphenoid base (Fig. 2.22). Craniofacial disassociation is completed centrally by fracture of the perpendicular plate of the ethmoid below the cribriform plate, and peripherally by zygomatic arch fractures (Fig. 2.23).

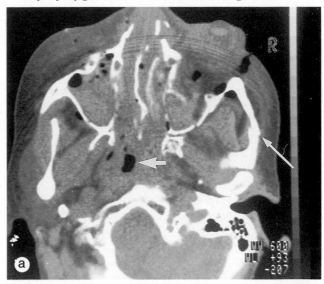

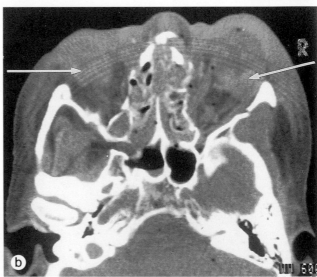

Fig. 2.23   Le Fort II–III fractures. Axial CT sections at lower orbital (**a**) and mid-orbital (**b**) level showing fractures of the nasoethmoidal region, medial, lateral and inferior orbital walls, zygomatic arch (long arrow in (**a**)), nasal septum, pterygoid plates and sphenoid sinus. Note also the bilateral globe ruptures with intraocular haematomas (arrows in (**b**)) and distortion of the upper nasopharyngeal airway (short arrow in (**a**)).

The above descriptions are of classical complete Le Fort fractures, with both sides of the face involved. Bilateral but incomplete forms also occur as can unilateral Le Fort fractures. Both sides of the face may be involved by Le Fort fractures at different levels and it may be necessary to describe each side of the face separately.

It should be appreciated that the fractures occurring along the Le Fort lines of weakness are usually comminuted and that the separated portion of mid-face is itself often fragmented, rather than one intact section of bone. The force required to produce a Le Fort fracture is considerable and therefore injury to adjacent parts of the face or intracranial cavity is common. A pure Le Fort fracture occurring along only one line of weakness is unusual. When seen, this is most commonly a Le Fort II fracture. Usually, Le Fort fractures occur in combination, for example, Le Fort I plus II (Fig. 2.20) or Le Fort II plus III (Figs 2.22 and 2.23). Associated zygomatic complex, orbital, hard palate, nasofrontal and nasoethmoidal fractures are commonly encountered [6].

The soft tissue complications which may be seen in these injuries are extensive, as multiple facial struts in all three orthogonal planes may be involved. Airway compromise from retropharyngeal haematoma formation must be excluded at the earliest opportunity, as fractures through the pterygoid plates (posterior coronal strut) are, by definition, present in Le Fort injuries.

PENETRATING INJURIES OF THE ORBIT

The soft tissue structures of the orbit are vulnerable to direct injury by sharp objects, such as pointed sticks, or projectile foreign bodies, including air- or BB-gun pellets and bullets. Intracranial penetration through the relatively thin orbital walls can easily occur even with seemingly innocuous objects. CT readily defines soft tissue injuries to the orbit including haematomas, globe disruption and optic nerve damage, as well as any associated intracranial complications. Metallic foreign bodies are easily located and usually cause streak artefact, but this does not detract significantly from the valuable soft tissue and bony information available (Fig. 2.24).

Fig. 2.24 Pellet (BB) gun injury of the left orbit. Axial CT sections through the mid- (**a**) and upper (**b**) orbit reveal multiple metal fragments which have breached the medial and supero-lateral orbital walls near the orbital apex, lodging within the surrounding paranasal sinuses and intracranial cavity. Note the haematoma behind the left globe in (**a**). Image distortion from metal artefacts is minimized by displaying the scans on a wide window.

MANDIBULAR FRACTURES

Fractures of the mandible may occur as a solitary injury or in association with other facial injuries. Plain radiographs and orthopantomograms normally depict fractures of the body, ramus and condylar region satisfactorily and CT is not usually necessary. On occasions, CT can be a useful complementary examination in complex mandibular injuries, for example, in cases of displaced condylar neck fractures (Fig. 2.25).

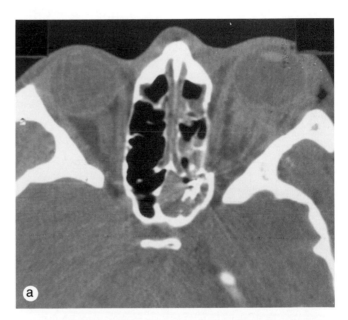

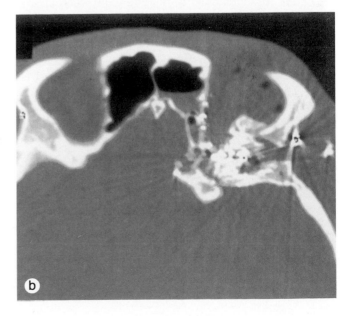

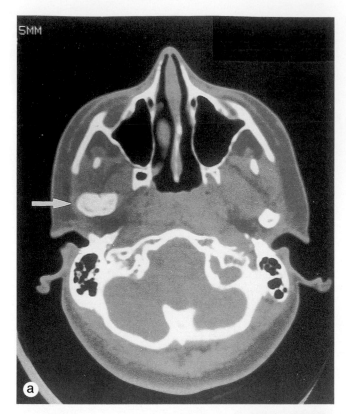

Fig. 2.25 Right condylar neck fracture (a) Axial CT section just below the level of the mandibular condylar fossae. The condylar head has been drawn anteriorly and inferiorly by the lateral pterygoid muscle and has fused with the condylar neck, giving the oval-shaped bony structure demonstrated on CT (arrow). (**b**) Lateral 3D CT image. The empty right condylar fossa is apparent and the distorted condylar head/neck mass has formed a pseudarthrosis with the articular eminence (arrows).

Fig. 2.26 Temporomandibular joint: magnetic resonance imaging. Sagittal-oblique T1-weighted spin-echo images (SE500/25). With the mouth closed (**a**) the posterior band of the meniscus lies in the 11 o'clock rather than the normal 12 o'clock position in relation to the condylar head, but is within normal limits. In the mouth open position (**b**) the meniscal/condylar head relationship is normal.

Most of the essential information is obtained from T1-weighted images in the sagittal or sagittal-oblique plane, at right angles to the long axis of the condylar head. Scans are performed with the mouth in the open and closed positions, with coronal images used to detect medial or lateral meniscal displacements. Both joints are usually studied as the incidence of bilateral abnormalities is high (50–80%).

Internal derangements of the temporomandibular joint may follow trauma to the jaw. MRI has significant advantages over CT for the visualization of the joint meniscus: soft tissue detail is superior, direct multiplanar imaging is considerably easier, and it avoids ionizing radiation near the eyes.

The temporomandibular joint meniscus is a biconcave, low-signal structure (on both T1- and T2-weighted sequences) with anterior and posterior bands at either end and a narrower intermediate zone in between [15]. Normally, the posterior band lies in the 12 o'clock position in relation to the condylar head. As the mouth opens, the meniscus translates forwards in unison with the condylar head as far as the temporal eminence (Fig. 2.26). Abnormalities of meniscal position and movement can be assessed (Fig. 2.27), as well as arthritic and in-

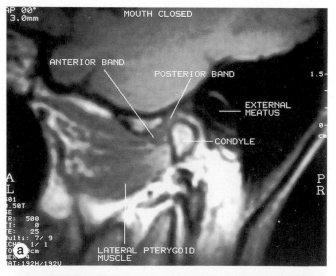

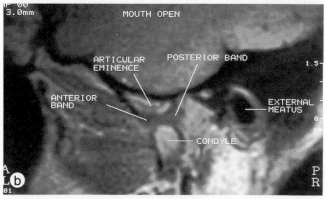

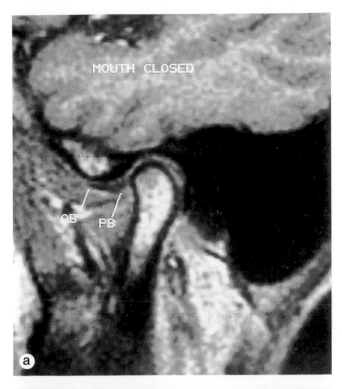

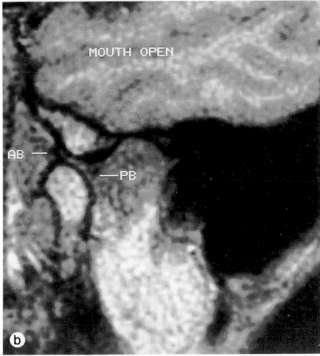

Fig. 2.27 Temperomandibular joint: internal derangement. Sagittal-oblique T1-weighted gradient echo images (GE300/14/90°). With the mouth closed (**a**) the meniscus is anteriorly dislocated, but with the mouth open (**b**) the normal relationship of the meniscus and condylar head is restored. In more severe cases, the dislocation does not reduce on mouth opening and meniscal deformity may be identified. AB = anterior band; PB = posterior band.

flammatory joint conditions. The latter may require additional T2-weighted images for evaluation. Kinematic MR studies of joint movement are becoming feasible with images obtained of the temporomandibular joint at multiple points during its movement cycle and displayed in a cine mode. A detailed discussion of this specialized subject is beyond the scope of this chapter.

MAGNETIC RESONANCE IMAGING IN FACIAL TRAUMA

At present, MRI has a limited though increasing role in the assessment of facial injury. The status of MRI in this area has recently been reviewed by Gentry [16]. The disadvantages of MRI compared to CT can be summarized as follows.

1. MRI is not usually practical in the severely injured or unco-operative patient, because of the longer scanning time required and the difficulty in bringing life-support equipment into the MRI environment.
2. Fractures are more difficult to detect because of the signal void of cortical bone, particularly when adjacent to low-signal air spaces, and the generally poorer spatial resolution of MRI.
3. If metallic foreign bodies are present around the orbit, then MRI is contraindicated. Non-metallic foreign bodies are also less easily seen than with CT.

T1-weighted sequences are the most useful, as they are faster to perform and have a higher signal-to-noise ratio than T2-weighted sequences.

Advantages of MR over CT are as follows:

1. Direct multiplanar imaging can be performed without moving the patient and without any radiation penalty.
2. Soft tissue contrast resolution is superior on MR and so soft tissue injury can be better characterized. Intraorbital and intraocular haematomas are detected, classified and aged with greater accuracy than CT. Optic nerve contusions and certain intracranial complications, such as axonal (shearing) injury or non-haemorrhagic contusions can be seen on MRI with greater sensitivity than CT, using T2-weighted sequences.

MRI can therefore provide important complementary information to CT in selected situations (Fig. 2.28). The drawbacks of MRI will preclude it

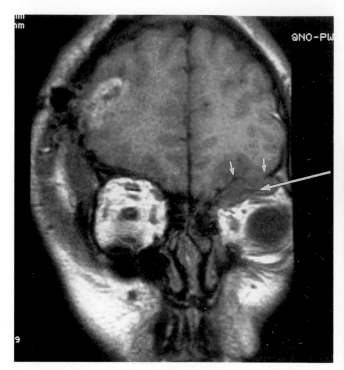

Fig. 2.28 Left orbital roof 'blow-in' fracture. Coronal T1-weighted gradient echo image (GE380/14/90°). The patient was assaulted 24 hours previously with a baseball bat, suffering a depressed fracture to the right calvarium with underlying haemorrhagic brain contusion. There were no external signs of left-sided craniofacial injury but CT disclosed a left orbital roof fracture with bone fragments and abnormal soft tissue in the upper orbit but no other left-sided fracture. MRI was undertaken to further evaluate the superior orbital soft tissue. The MR scan confirms the soft tissue to be oedematous brain herniating through the superior orbital defect (fracture margins denoted by small white arrows). The depressed bone fragment is shown as an area of low signal (long white arrow). The areas of high signal in the right cerebral hemisphere represent methaemoglobin within the cerebral contusion. Debridement of the cranial injury and elevation of the depressed fracture had already been performed.

becoming the imaging investigation of choice over CT in most types of maxillofacial injury for the forseeable future. However the area where MRI may soon present a serious challenge to CT is in the assessment of blow-out fractures [17]. On T1-weighted images the location of fractures can be inferred from herniation of high-signal intraorbital fat into the adjacent paranasal sinuses, even if the fractures themselves are not reliably depicted. Lower-signal structures, including acute haematomas (containing deoxyhaemoglobin), rectus muscles, sinus fluid and soft tissue adhesions contrast well within the high-signal fat. Assessment of motility disorders is facilitated by the multiplanar imaging capability, allowing, in particular, sagittal-oblique images of the

inferior rectus muscle region (Fig. 2.29). T1-weighted images thus provide most of the information needed in acute blow-out injuries. Subacute injuries may require additional T2-weighted sequences to help differentiate methaemoglobin-containing haematomas from fat, both of which have high signal on T1-weighted images.

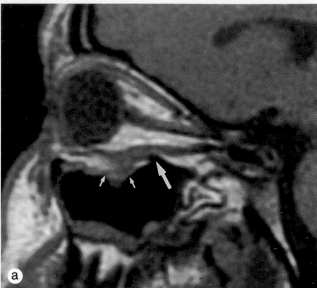

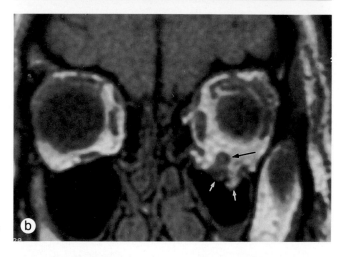

Fig. 2.29 Orbital floor blow-out fracture: magnetic resonance imaging. (**a**) Sagittal-oblique T1-weighted spin-echo image (SE500/25). An extensive orbital floor fracture is evident with downward herniation of orbital contents (small white arrows). The herniated contents represent high-signal orbital fat and heterogeneous signal areas extending downwards from the inferior rectus muscle, representing areas of haemorrhage and local oedema. The inferior rectus muscle itself is swollen and slightly oedematous and looks to be impinged on by the posterior lip of the floor fracture (large white arrow). (**b**) Coronal T1-weighted spin-echo image (SE520/25). The herniated orbital contents are indicated (small white arrows) as well as the swollen inferior rectus muscle (black arrow).

ACKNOWLEDGMENTS

I would like to thank Olwen Guthrie for typing this manuscript and Mr John Yates for producing the illustrations.

REFERENCES

1. Beck RA, Blakeslee DB. The changing picture of facial fractures. Archives in Otolaryngology and Head and Neck Surgery 1989; 115: 826–829.
2. MacEwen CJ, Jones NP. Eye injuries in racquet sports. British Medical Journal 1991; 302: 1415–1416.
3. Manson PN, Markowitz B, Mirvis S, Dunham M, Yaremchuk M. Toward CT-based facial fracture treatment. Plastic and Reconstructive Surgery 1990; 85(2): 202–212.
4. Gentry LR, Manor WF, Turski PA, Strother CM. High-resolution CT analysis of facial struts in trauma: 1. Normal anatomy. American Journal of Roentgenology 1983; 140: 523–532.
5. Gentry LR, Manor WF, Turski PA, Strother CM. High-resolution CT analysis of facial struts in trauma: 2. Osseous and soft tissue complications. American Journal of Roentgenology 1983; 140: 533–541.
6. DeLaPaz R, Brant-Zawadzki M, Row LD. CT of maxillofacial injury. In: Federle MP, Brant-Zawadzki M, eds, Computed tomography in the evaluation of trauma. Baltimore, MD: Williams and Wilkins, 1986: p. 64–107.
7. Jackson A, Whitehouse RW. Low dose computed tomography imaging in orbital trauma. British Journal of Radiology (in press).
8. Koorneef L, Zonneveld FW. The role of direct multiplanar high resolution CT in the assessment and management of orbital trauma. Radiologic Clinics of North America 1987; 25(4): 753–766.
9. Gillespie JE, Isherwood I, Barker GR, Quayle AA. 3-Dimensional reformations of computed tomography in the assessment of facial trauma. Clinical Radiology 1987; 38: 523–526.
10. Gholkar A, Gillespie JE, Hart CW, Mott D, Isherwood I. Dynamic low-dose 3-dimensional computed tomography: a preliminary study. British Journal of Radiology 1988; 61: 1095–1099.
11. Gentry LR, Smoker WRK. Computed tomography of facial trauma. Seminars in Ultrasound, CT and MR 1985; 6(2): 129–145.
12. Manson PN, Ruos EJ, Iliff NT. Deep orbital reconstruction for correction of post-traumatic enophthalmos. Clinics in Plastic Surgery 1987; 14 (1): 113–121.
13. Whitehouse RW, Jackson A. Measurements of orbital volumes following trauma using low-dose computed tomography. European Radiology 1993; 3: 145–149.
14. Hammerschlag SB, Hughes S, O'Reilly GV, Naheedy MH, Rumbaugh CL. Blow-out fractures of the orbit: a comparison of computed tomography and conventional radiography with anatomical correlation. Radiology 1982; 143: 487–492.
15. Fulmer JM. Temporomandibular joint. In: Stark DD, Bradley GG, eds, Magnetic resonance imaging, 2nd ed. St Louis, MO: CV Mosby/Year Book, 1992: p 1244–1267.
16. Gentry LR. Facial trauma and associated brain damage. Radiologic Clinics of North America 1989; 27(2): 435–446.
17. Tonami H, Yamamoto I, Matsuda M et al. Orbital fractures: surface coil MR imaging. Radiology 1991; 179: 789–794.

3 The skull base

T. Jaspan and I. Holland

The skull base represents the bone mass lying between the facial skeleton, its contents and the three cranial fossae. This complex structure functions not only as a protective floor and source of anchorage for the delicate intracranial contents, but also as a conduit for the passage of numerous neurovascular elements in both directions.

ANATOMY

The skull base, as viewed by the neurosurgeon, forms the base to his 'territory', as opposed to the otolaryngologist who views the skull from below. In general, the neurosurgical anatomy is divided into three tiered plateaus — the anterior, middle and posterior cranial fossae. To the otolaryngologist the skull base consists of an anterior component, including the midline structures extending from the frontal sinus to the anterior margin of the foramen magnum, and a lateral part formed by the paired petrous temporal bones. For the purpose of this review, the skull base has been divided into anterior and postero-lateral compartments.

Anterior skull base

The rostral extent is formed by the frontal sinus. The roof of the ethmoid sinus with the midline crista galli and the paper-thin cribriform plate lies immediately posteriorly. The latter is continuous with the planum sphenoidale and the tuberculum sellae. Anterior to the crista galli lies the foramen caecum, between the frontal and ethmoid bones, usually vestigial in the adult. There are no major canals or foramina at this level. Laterally the floor is formed by the orbital plates of the frontal bone anteriorly and the lesser wing of the sphenoid bone posteriorly. The bony structures are best appreciated by direct coronal plane CT scanning (Fig. 3.1); however MR imaging in the sagittal plane elegantly displays the relationships of the underlying nasal cavity, soft tissue elements and paranasal sinuses to the bony skull base (Fig. 3.2).

Fig. 3.2 Midline T1-weighted MRI. The clivus (1) is separated from the nasopharynx inferiorly by adenoidal tissues and the nasopharyngeal mucosa and submucosa. The middle (2) and inferior (3) turbinates lie anteriorly within the nasal cavity, separated from the oral cavity by the hard (4) and soft (5) palate. Note the void signal from the sphenoid sinus (open arrow), high signal from the marrow contained within the crista galli (curved arrow) and the sphenoethmoidal recess (white arrow).

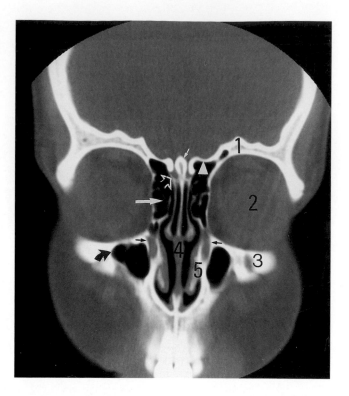

Fig. 3.1 Coronal CT scan through the anterior skull base. The orbital plate of the frontal bone forms the roof of the orbit (1) with the contained globe (2) inferiorly. The inferior orbital nerve canal (3) is seen within the floor of the orbit, the latter separating the orbit from the underlying maxillary sinus (curved arrow). The nasal septum (4) and inferior turbinate (5) are seen in the midline. Note the nasolacrimal ducts (small black arrows), cribriform plates (open white arrow) lying caudal to the roof of the ethmoid air cells (white arrowhead), the ethmoid air cells (long white arrow) and the midline bony projection of the crista galli (small white arrow).

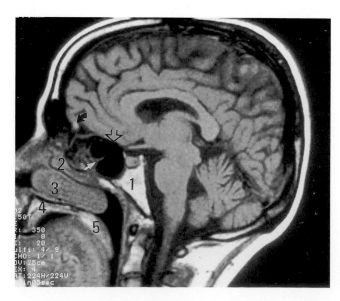

Passing posteriorly, the body of the sphenoid bone forms the centre of a butterfly-shaped bone mass. The central body houses the sella turcica and sphenoid sinus with the cavernous sinuses lying laterally. Meckel's cave containing the trigeminal ganglion grooves the floor medially. Extending laterally, the greater wing of the sphenoid bone forms the relatively thin anterior wall of the middle cranial fossa in the form of the vertically orientated orbital plates. These sweep postero-inferiorly, becoming thicker and uniting with the petrous temporal bones to form the floor laterally.

Contained within this bony superstructure are numerous fissures and foramina. The superior orbital fissure lying between the greater and lesser wing of the sphenoid bone connects the intracranial compartment with the orbit and transmits cranial nerves (CN) III, IV, VI and the first division of V along with the superior ophthalmic vein and middle meningeal artery branches. There is also communication with the pterygopalatine fossa via the inferior orbital fissure (Fig. 3.3). The second division of the fifth

cranial nerve passes into the pterygopalatine fossa via the foramen rotundum.

Posterior to this the foramen ovale extends inferiorly, as best appreciated on coronal scanning, to communicate with the infratemporal fossa, transmitting the mandibular branch of the fifth cranial nerve and emissary veins connecting the cavernous sinus with the pterygoid plexus of veins (Fig. 3.4).

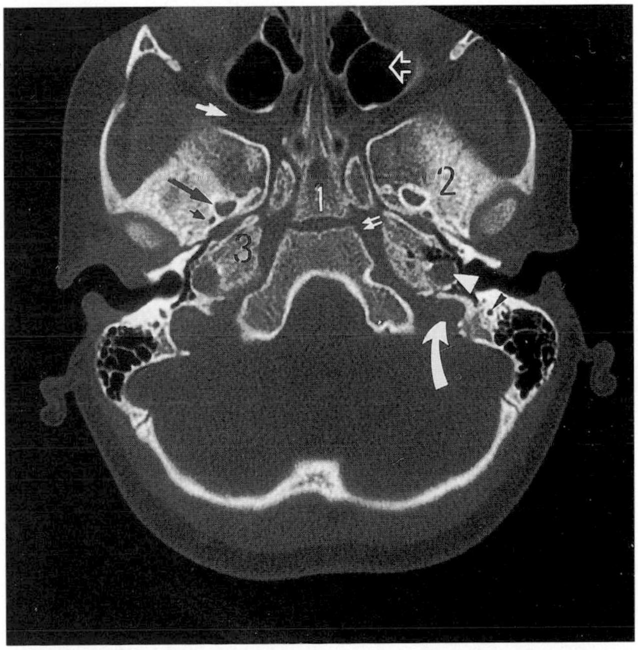

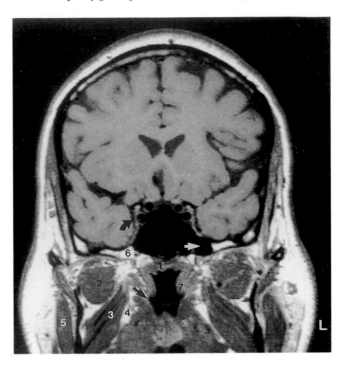

Fig. 3.4 T1-weighted MR scan at the level of the foramen ovale, infratemporal fossa and nasopharynx. The sphenoid sinus lies above the nasopharynx, separated by the longus capitus muscle belly (1). The lateral (2) and medial (3) pterygoid muscles are easily recognized, lying lateral to the hyperintense fat within the parapharyngeal space (4). The masseter muscle (5) lies lateral to the mandible. High signal is also seen from fat in the marrow of the sphenoid bone (6). Note the torus tubarius (7) with the orifice for the eustachian tube below (black arrow). A prominent lateral recess of the sphenoid sinus is present (white arrow). The cavernous sinus (curved black arrow) contains flow voids from the contained internal carotid artery.

Fig. 3.3 Axial CT scan at the level of the pterygopalatine fossa. The maxillary sinus (open arrow) is separated from the greater wing of the sphenoid bone (2) by the pterygopalatine fossa (white arrow). Note the clivus (1), petrous apex (3), foramen ovale (long black arrow), foramen spinosum (short black arrow), foramen lacerum (double white arrows), carotid canal (white arrowhead), jugular foramen (curved white arrow) and facial nerve canal (black arrowhead).

The middle meningeal artery and vein pass through the foramen spinosum lying more laterally, with the fibrocartilage-covered foramen lacerum lying medially. The latter provides further potential communication between the middle fossa and the nasopharynx below. The internal carotid artery (ICA) enters the skull through the petrous bone medial to the styloid process and then turns anteromedially to enter the carotid canal, located within the apex of the petrous bone.

It then passes above the foramen lacerum with the bony canal for the eustachian tube lying immediately laterally. Passing superiorly and medial to the trigeminal ganglion, the ICA enters the cavernous sinus.

The posterior aspect of the anterior skull base is formed by the dorsum sellae and clivus, the latter forming the anterior margin of the foramen magnum. The foramen magnum contains the medulla oblongata, the spinal portion of CN XI, the vertebral arteries and veins and the anterior and posterior spinal arteries. The hypoglossal canal lies medially, situated between the occipital condyles and the jugular tubercule. As well as the hypoglossal nerve it transmits a meningeal branch of the ascending pharyngeal artery.

Posterolateral skull base

The petrous bones form the anterolateral margins and, together with the occipital bone, the concave inner surfaces for the cerebellar hemispheres. The transverse and superior petrosal sinuses groove the petrous pyramids. The sigmoid sinus descends via a groove indenting the mastoid bone and jugular process of the occipital bone to form the jugular bulb within the jugular foramen. Also transmitted through the foramen are cranial nerves IX, X and XI and meningeal branches of the ascending pharyngeal and occipital arteries. The jugular foramen extends superomedially to the petro-occipital fissure, containing the inferior petrosal sinus which connects the cavernous sinus with the jugular bulb. The posterior aspect of the petrous bone contains the internal auditory canals, carrying cranial nerves VII and VIII, and the vestibular and cochlear aqueducts. Axial plane scanning provides excellent topographical delineation of these various foramina while coronal imaging is particularly useful in assessing the relationships of the foramen magnum, jugular bulbs and the stylomastoid foramina, through which CN VII exits. The labyrinthine structures, middle ear clefts and external auditory canals are often best assessed by a combination of axial and coronal plane scanning (Fig. 3.5).

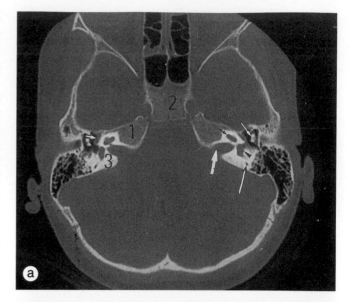
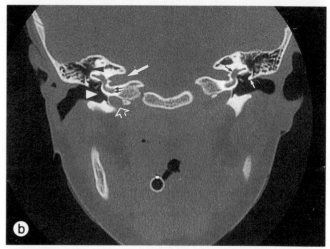
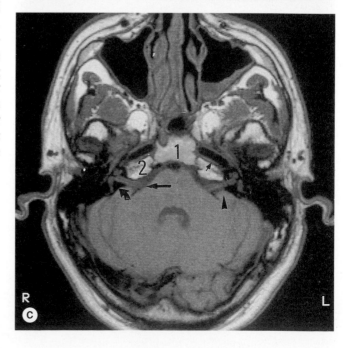

Fig. 3.5 Axial and coronal CT (a), (b) and axial T1-weighted MR (c) images through the labyrinthine structures. (a) The petrous apex (1), clivus (2) and otic capsule (3) are the principal bony landmarks. The vestibule (black arrowhead) and apical turn of the cochlea (small black arrow) are readily identified within the otic capsule. The tympanic section of the facial nerve canal may be seen lying lateral to the cochlea (white arrow) with the ossicles (long thin white arrow) contained within the middle ear cleft. Note the internal auditory canal (long thick white arrow) and

posterior semicircular canal (long thin white arrow). (**b**) The internal auditory canal (thick white arrow), vestibule (short black arrow), basal turn of cochlea (double black arrows), attic (black arrowhead), scutum (small white arrow) and middle ear cleft (white arrowhead) are well displayed in the coronal projection. Note the relationship of the carotid canal (open white arrow) to the hypotympanum.
(**c**) Hyperintensity is seen from marrow fat in the clivus (1) and within the petrous apex (2), the seventh and eighth nerves are clearly seen (long arrow) traversing the cerebellopontine angle cistern.

The flocculus of the cerebellum (arrowhead) abuts against the posterior aspect of the traversing nerves. Note the cochlea and vestibule surrounded by the void signal of the otic capsule (curved arrow). A signal void from the internal carotid artery is seen within the carotid canal (short arrow).

IMAGING TECHNIQUE

As the major component forming the skull base is bone, CT is particularly useful in the assessment of the radiological anatomy of this region. The development of magnetic resonance imaging provides greater scope for the display of soft tissue elements of the skull base and facial planes, enabling superior assessment of pathological processes. These two modalities are complementary in many ways. The selection of either is dependent upon local access and specific requirements; however we shall consider both together.

Computed tomography

Demonstration of structures in and contiguous to the skull base requires axial scanning with gantry tilt approximately 20° to the orbito-meatal plane. Depending on the area to be covered and detail required, 3–5 mm-thick slices are employed. Images should be interrogated with a wide window (> 1600 HU) for visualization of bone and air-containing structures and basal foramina. A narrow window (> 400 HU) is used for soft tissue detail, providing greater contrast.

Intravenous contrast material is administered for identification of vessels, vascular tumours, anomalies and infective processes. Coronal scanning is performed perpendicular to the orbito-meatal plane or approximately 90° to the hard palate and may be performed in the hyperextended supine or prone position, depending on the patient's tolerance of either position. There are certain circumstances which necessitate scanning in the prone position, particularly the investigation of suspected cerebrospinal fluid fistulae and paranasal sinus inflammatory disease.

The petrous temporal bone and contents are optimally studied employing 1–2 mm-thick slices employing special bone algorithm protocols. Scanning is performed in both axial and coronal planes. The wide windows employed rule out the use of contrast medium.

Magnetic resonance imaging

The multiplanar capabilities of MR imaging are of considerable value in the assessment of diseases affecting the skull base. Coronal plane imaging is particularly useful for defining the inferior or superior extent of lesions involving the cavernous sinus and sella. The axial plane is useful for assessing the spaces immediately below the skull base, specifically the parapharyngeal spaces, pterygopalatine fossae and lateral pharyngeal recesses (fossae of Rosenmuller). Midline sagittal imaging is especially helpful in assessing disease involving the paranasal sinuses, clivus and pituitary fossa as well as midline congenital defects.

T1- and T2-weighted spin-echo or gradient-echo pulse sequences are used employing 3-5 mm-thick slices in most areas. The T1-weighted (T1WI) image gives superior soft tissue and anatomical detail. On T2-weighted (T2WI) images, pathological processes tend to have a brighter signal than normal tissues, enabling their easier identification.

Some lesions, notably squamous cell carcinomas, may however appear isointense or even of reduced signal intensity with respect to adjacent normal structures, making them less conspicuous. Low FLIP angle gradient echo pulse sequences are employed for the demonstration of blood vessels and are the most sensitive for the detection of calcifications. MR imaging is however inferior to CT scanning for the demonstration of fine bony structures, calcifications and subtle modifications of the normal bony architecture.

The recent introduction of MR contrast agents, used especially in conjunction with gradient echo or fast spin-echo techniques, offers the optimal imaging technique for the assessment of inflammatory and neoplastic conditions in the future. A note of caution should be sounded, however, in relation to the use of contrast enhancement in this region. Short TR/TE images rely heavily on the tissue contrast provided by normal fat in the soft tissues underlying the skull base to delineate the full extent of disease involvement. Enhancement within lesions may in some cases actually obscure their presence unless fat suppression techniques are employed.

PATHOLOGY – ANTERIOR SKULL BASE

Congenital

Congenital lesions involving the skull base may be considered as arising either within the cranial (endocranial) or bony compartments.

Endocranial lesions

Encephaloceles

These abnormalities are characterized by extracranial herniation of brain tissues and meninges through a defect in the dura and skull base. The commonest variety is the fronto-nasoethmoidal encephalocele followed by sphenoidal lesions [1,2]. Imaging is intended to define the contents of the protrusion, specifically the presence of brain within the sac, and the incorporation of any vascular elements within the protrusion. The morphological appearance of the residual brain needs to be assessed and a search must be made for any associated anomalies such as agenesis of the corpus callosum, which occurs in approximately 80% of cases [3].

While the anterior basal encephaloceles commonly present at birth with an obvious midline soft pulsatile swelling, sphenoidal encephaloceles may escape detection until later in the first or second decade, when the presentation is typically one of nasal obstruction produced by the herniated sac filling the nasopharynx. CT scanning may demonstrate a sclerotic margin at the site of the defect as well as the soft tissue contents of the protrusion (Fig. 3.6). Coronal scanning may define the anatomy and relationships of the lesion, but MR imaging offers the most complete assessment with clear definition of the dysplastic brain extending into the sac, the surrounding cerebrospinal fluid (CSF) and associated vessels.

The sagittal plane is also important to show the location of the optic pathways and pituitary infundibulum, which may become incorporated into the sac.

Arachnoid cysts

These are CSF-filled collections which are in most cases due to anomalous development of the arachnoid membrane [4]. They are most commonly located in the middle fossa where there is often hypoplasia of the underlying temporal lobe (Fig. 3.7). Chronic CSF pulsations produce localized pressure erosion and expansion of the bone which is well demonstrated by CT scanning.

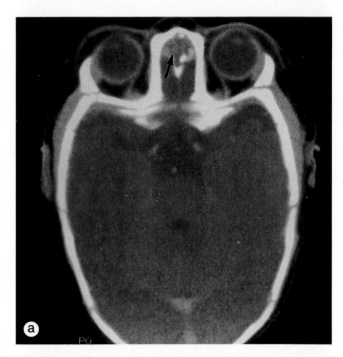

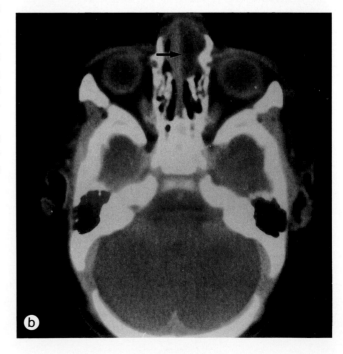

Fig. 3.6 Frontonasal meningoencephalocoele demonstrated on axial CT scanning. Note the soft tissue mass (arrow) extending inferiorly into the nasal cavity through the bony defect. The low density of the lesion suggests that the sac contains mainly cerebrospinal fluid.

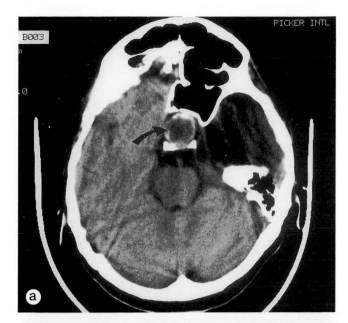

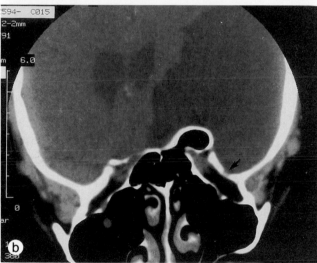

Fig. 3.7 Axial (a) and coronal (b) CT scans demonstrating a large middle cranial fossa arachnoid cyst. There is focal thinning of the bony floor (arrow). Note the presence of associated pneumatosis dilatans with pneumatization of the anterior clinoid process. A craniopharyngioma is also present (curved arrow).

Dermoid and epidermoid cysts

Extracranial dermoid cysts are congenital lesions resulting from ectodermal inclusions into the closing neural groove or embryonal neurectodermal projections, such as through the foramen caecum anteriorly. These defects probably occur during the third to fifth week of embryogenesis. These classically midline lesions are embryologically related to the encephaloceles and nasal gliomas. Dermal and epidermal elements are characteristically found and there may be evidence of a persistent track or

communication with the skin, as in the case of the nasal dermoid. If an intracranial extension exists there is a risk of meningitis [5].

CT scanning demonstrates the lesions to be of low attenuation. High resolution scanning in the coronal plane is recommended in the assessment of nasal dermoids, with particular emphasis on the presence or absence of intracranial extension. The latter is suggested by a bifid crista galli and a patent foramen caecum [6].

MRI, however, provides superior morphological information and better tissue contrast. The sagittal plane enables visualization of the cyst in its entirety. The cysts are characteristically hyperintense on short TR/TE MR images, reflecting the high concentration of fatty material and are of corresponding low intensity on long TR/TE sequences. In the context of a nasal mass, MRI is the optimal modality for differentiating the lesion from a frontonasal encephalocele [7]. Intracranial dermoid cysts may occasionally arise from the suprasellar region and extend inferiorly to involve the skull base. These cysts may rupture into the subarachnoid space.

Although often acquired, epidermoid cysts presenting in early life are usually congenital in origin, representing incomplete cleavage of neural from cutaneous ectoderm. The cysts are squamous-epithelium-lined and grow by epithelial desquamation, the cells being converted to keratin and cholesterol crystals. The CT appearances are of a low density midline lesion which may exhibit peripheral calcification (Fig. 3.8). On MR imaging the cysts are

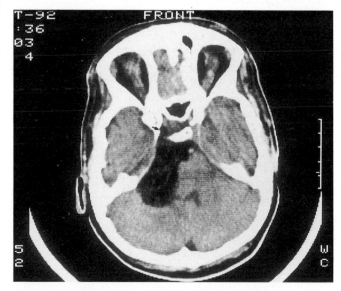

Fig. 3.8 Axial CT scan demonstrating a large epidermoid cyst based upon the right cerebellopontine angle cistern. The lesion extends through the incisura into the parasellar region (arrow) and insinuates itself around the brain stem anteriorly.

usually of inhomogeneous low intensity on T1-weighted images (T1WI) and hyperintense on T2-weighted images (T2WI) [8] (Fig. 3.9). Proton density imaging enables differentiation from arachnoid cysts, which they may closely resemble, as epidermoid cysts do not appear isointense with respect to cerebrospinal fluid, unlike arachnoid cysts.

Chiari malformation

First described in 1891 [9], these lesions are characterized by malformations of the hindbrain associated with varying degrees of caudal displacement of structures through the foramen magnum, frequently accompanied by hydrocephalus. The Chiari I malformation occurs probably as a result of craniocervical dysgenesis, not infrequently associated with skull base and upper cervical anomalies such as the Klippel–Feil complex and occipitodental fusion or consequential upon acquired abnormalities such as basilar invagination and platybasia. Normal cerebellar tonsils may protrude up to 3 mm below a line joining the opisthion to the basion, typically with a rounded configuration of the tonsils. Cerebellar ectopia exceeding 5 mm, with a characteristically pointed configuration of the tonsils, is diagnostic of a Chiari I lesion. Increasing degrees of herniation of tonsils are accompanied by rapidly increased frequency of clinical symptoms.

Disturbed cerebrospinal fluid flow patterns at the level of the foramen magnum may account for the 20–25% incidence of syringohydromyelia. MR imaging in the sagittal plane displays extremely well the anatomy and associated cord and/or brainstem cavitation (Fig. 3.10).

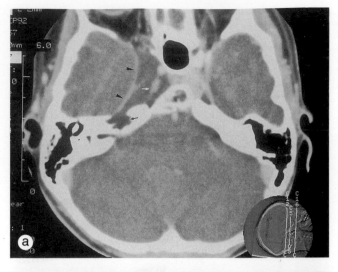

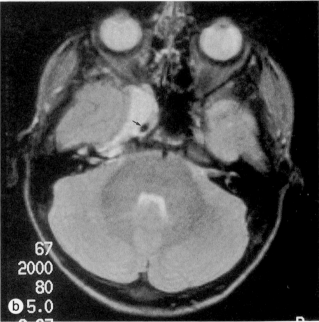

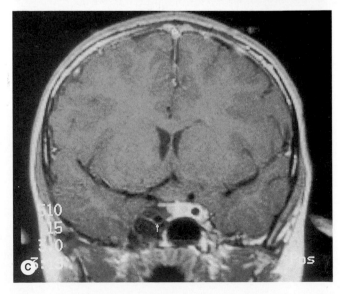

Fig. 3.9 Epidermoid cyst in an 11-year-old male presenting with recurrent meningitis. (a) Axial CT scan showing a homogeneously low attenuation lesion within the right cavernous sinus encasing the traversing internal carotid artery (white arrow). The intact dura (arrowheads) separates the tumour from the temporal lobe. Note the involvement the adjacent petrous apex (black arrow). **(b)** Axial T2-weighted image demonstrates a high signal intensity lesion in the right parasellar region encasing the internal carotid artery (arrows), extending along the medial surface of the middle cranial fossa and encroaching upon the orbital apex. **(c)** Coronal enhanced T1-weighted images showing the lobulated configuration of the lesion which does not enhance. Note again the encased internal carotid artery (arrow).

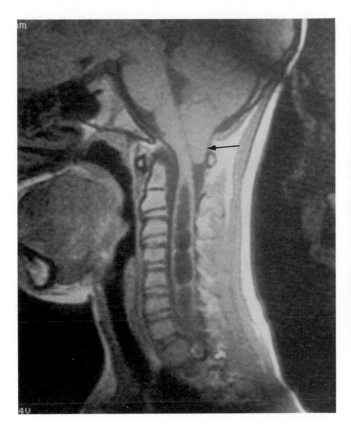

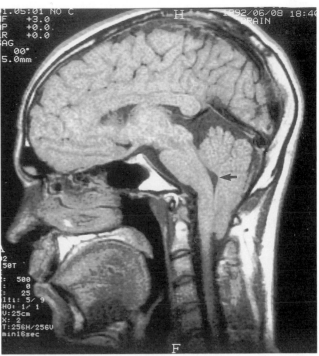

Fig. 3.10 Chiari I malformation. Sagittal T1-weighted image showing cerebellar tonsillar herniation through the foramen magnum (arrow). The fourth ventricle has a normal configuration and position. Note the large hydrosyrinx involving the cervical cord with pseudoseptations internally.

Fig. 3.11 Chiari II malformation. Sagittal T1-weighted image demonstrating the characteristic shallow posterior fossa and herniation of hindbrain structures through the foramen magnum. The fourth ventricle is barely visualized as a slit-like low signal intensity (arrow) extending abnormally caudad. A repaired cervicodorsal myelomeningocele is present with an adjacent cord cavity (arrowheads), confirmed on axial images.

The Chiari II malformation occurs uniquely in association with myelomeningoceles. Following closure of these lesions at birth, hydrocephalus is virtually inevitable. The precise pathogenesis is as yet incompletely understood but may well represent the end result of impaired development of the ventricular system leading to multiple anomalies of the developing brain. These complex lesions are characterized by a shallow posterior fossa, probably resulting in the cephalocaudal displacement of the contents [10]. There is a spectrum of abnormalities ranging from mild cerebellar ectopia to near total absence of the cerebellar hemispheres, the latter classified by some as a Chiari IV malformation. Inferior displacement of the cerebellar vermis, medulla and fourth ventricle may be accompanied by buckling of the medulla upon the relatively fixed upper cervical cord (Fig. 3.11). The fourth ventricle may appear almost normal, but will usually be slit like or occasionally ballooned as it becomes trapped by aqueduct stenosis above and basal cisternal compression below.

There is a high incidence of associated syringohydromyelia which may occur in the cervical cord or be more remote. A wide range of anomalies occurs in the supratentorial compartment in most cases [11]; MR imaging of the whole spine and intracranial compartment is once again the ideal imaging stratagem [12], particularly in assessing the presence or progression of cord cavitation. Chiari III malformations are rare abnormalities characterized by herniation of the hindbrain into a low occipital or high cervical encephalocele associated with the other pathological and radiological features of a Chiari II malformation [13].

Osseous lesions

Congenital craniovertebral junction anomalies

Included in this group is a variety of abnormalities which may occur in isolation or association with syndrome complexes such as the Klippel–Feil anomaly. Atlanto-occipital fusion, platybasia (a basal angle in excess of 142° to 144°) and basilar invagination may lead to varying degrees of encroachment of the structures at the foramen magnum.

Mid-sagittal MR is of particular value in establishing the relationships of the osseous and neural elements (Fig. 3.12), but high resolution CT with reformation or pluridirectional tomography can often provide most information about the complex bony anomalies.

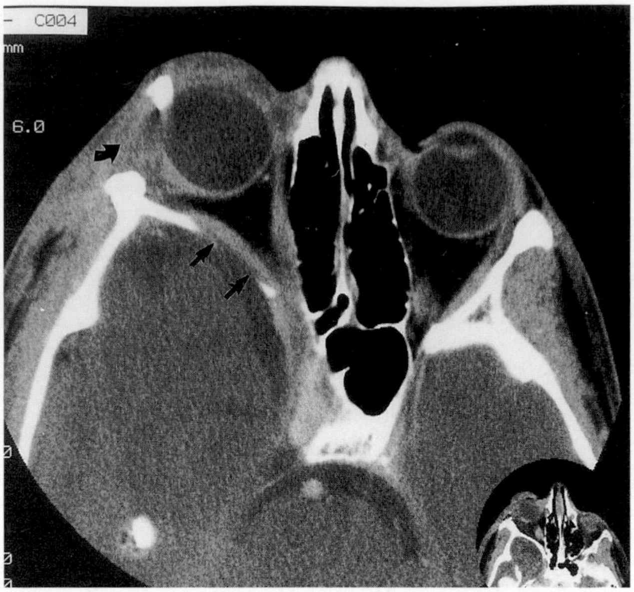

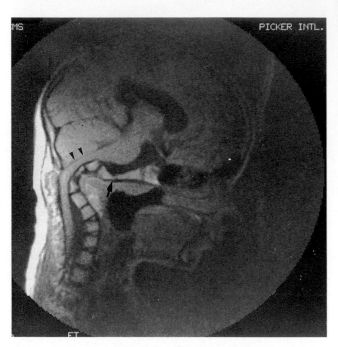

Fig. 3.12 Craniovertebral junction anomaly. Midline sagittal T1-weighted images demonstrating severe basilar invagination associated with platybasia, fusion and assimilation of the upper cervical vertebrae and occipito-dental fusion (arrow). There is marked medullary compression (arrowheads).

Fig. 3.13 Twenty-seven-year-old male with known neurofibromatosis. Axial CT scans showing sphenoid bone hypoplasia with a focal defect in the lateral wall of the right orbit (arrows), an expanded middle fossa floor and dysplasia of the temporal lobe. A plexiform neurofibroma is seen overlying the right orbit (curved arrow). An upper eyelid metallic implant is present.

Dysplasias

Dysplasia of the skull base is a particular feature of type 1 neurofibromatosis. The sphenoid bone, particularly the greater wing, is usually affected, with enlargement of the middle fossa. The combination of sphenoid bone, bony orbit dysplasia and buphthalmos with a plexiform neurofibroma is a characteristic of this type of neurofibromatosis (Fig. 3.13).

Hypoplasia

Achondroplasia is frequently associated with skull base hypoplasia. There is often stenosis of the foramen magnum, with compression of the medulla or upper cord and jugular foramina. Compression of the jugular veins leads to venous hypertension and consequently a CSF absorption block resulting in varying degrees of hydrocephalus occurring in virtually all patients. Skull base hypoplasia may also be seen as a part of cleidocranial dysplasia and craniofacial dysostosis (Crouzon's disease).

Basilar invagination

While this may be a feature of a skull base congenital anomaly or a congenital bony disorder such as osteogenesis imperfecta, more often it represents an acquired phenomenon whereby softening of the skull base leads to inversion of the foramen magnum margins with consequential stenosis and compression of the brainstem. Paget's disease, osteomalacia, hyperparathyroidism and osteoporosis associated with arthropathies such as rheumatoid arthritis may lead to varying degrees of compromise of the foramen magnum (Fig. 3.14).

Fibro-osseous lesions

This group of abnormalities is characterized by the replacement of normal bone by fibrous tissue. Lesions may be extensive and often symptomatic, either in terms of neurological/functional disturbance or a cosmetic deficit, or purely localized and frequently asymptomatic. This group includes the fibrous dysplasias, monostotic or polyostotic. Monostotic fibrous dysplasia particularly involves the maxilla, presenting in the first or second decades as a painless facial swelling. Other bones of the facial skeleton and skull base are also frequently involved, the lesions often being large. Plain radiographs typically show a 'ground glass' appearance of the lesion which blends imperceptibly with the surrounding normal bone. CT scanning is particularly useful in demonstrating the extent of the bony involvement and compression and narrowing of foramina and structures such as the orbit (Fig. 3.15) which may presage the need for decompressive surgery. Not infrequently, however, the lesions are slow growing and stabilize later in life. Malignant transformation is rare except in the context of previous radiotherapy, the latter now being contraindicated in the management of this condition. Polyostotic fibrous dysplasia less commonly involves the skull, the involvement being particularly of the skull base and occiput.

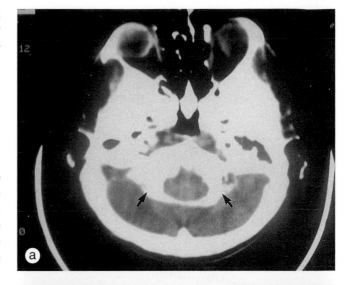

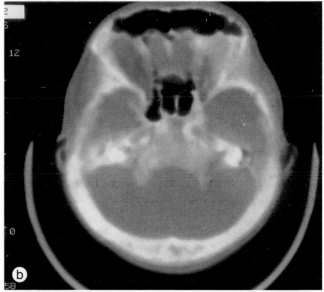

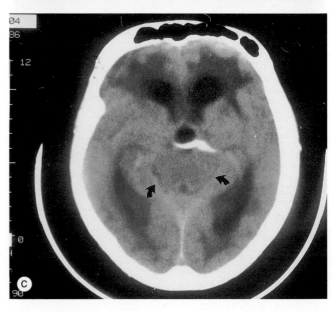

Fig. 3.14 Paget's disease. Axial CT scanning.
(**a**) Eversion of the margins of the foramen magnum which project into the posterior fossa (arrows). (**b**) Bone window settings show the expanded sclerotic skull base.
(**c**) Section at the level of the incisura shows effacement of the perimesencephalic cisterns (curved arrows) and an obstructive hydrocephalus with periventricular low attenuation.

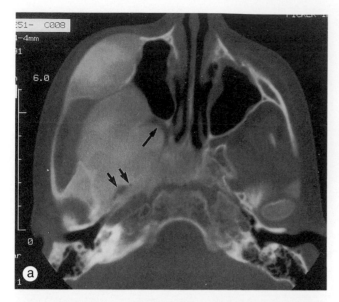

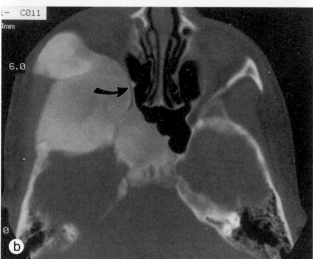

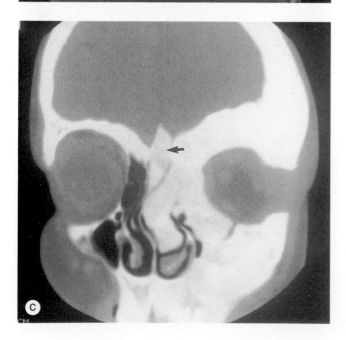

Tumours

Tumours arising within the cranial cavity

Tumours of neural origin

Cranial nerve tumours involving the anterior skull base are far less common than those involving the lateral skull base. Trigeminal and hypoglossal neurinomas are the main anterior skull-base lesions and characteristically erode the respective foramina producing, usually, a sclerotic rim on CT scanning as an indicator of their chronicity. Loss of definition or frank erosion should alert one to the possibility of malignant transformation [14].

The intracranial portions of the tumours appear isodense on CT scanning and demonstrate moderately intense enhancement following contrast administration. The component of the lesion penetrating the skull base and beyond does not appear to show such prominent enhancement. MR scanning is particularly useful in defining the skull-base involvement. Fifth nerve tumours may extend from the region of the Gasserian ganglion through the foramen ovale into the infratemporal fossa (Fig. 3.16).

Meningiomas

These tumours arise from meningothelial arachnoid cells. They are the most common primary intracranial neoplasm, representing approximately 15% of the total. More uncommonly, these tumours may arise from arachnoid cells originating in the internal auditory canal, geniculate ganglion, jugular foramen and the groove for the greater superficial petrosal nerve. Rarely, meningiomas may be entirely extracranial, probably related to ectopic arachnoid cell nests.

In general the different histological variants behave in a similar way apart from the locally aggressive angioblastic type, considered by some to be a haemangiopericytoma of the meninges. Presentation is commonly in the middle-aged patient, more often female.

Fig. 3.15 Fibrous dysplasia. Axial CT scans (**a**), (**b**) in a female with fibrous dysplasia demonstrating diffuse sclerotic expansion of the sphenoid and zygomatic bones. Note narrowing of the foramina ovalis and spinosa (short arrows in (**a**)), encroachment of the pterygopalatine fissure (long arrow in (**a**)) and almost complete obliteration of the inferior orbital fissure (curved arrow in (**b**)), maxillary sinus and narrowing of the orbit. Note the thickened crista galli on the coronal scan (**c**).

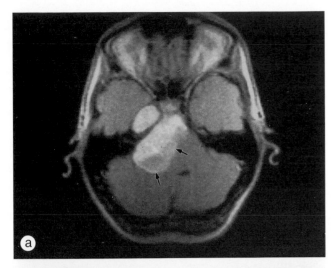

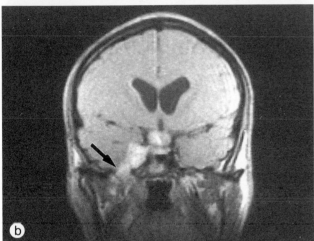

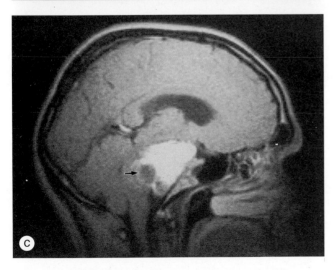

Of the intracranial skull-base tumours the commonest sites are the olfactory groove, planum sphenoidale, tuberculum sellae, parasellar region and middle fossa, clivus cerebellopontine angle and foramen magnum. The tumours are typically either encapsulated ovoid masses or more poorly delineated lesions extending along the dural surfaces in an 'en plaque' fashion. CT scanning typically shows a hyperdense mass which may contain amorphous calcifications. The lesion may also appear as a densely calcified mass (Fig. 3.17). The enhancement pattern is variable but usually tends to be strong

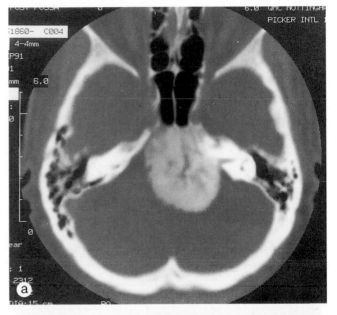

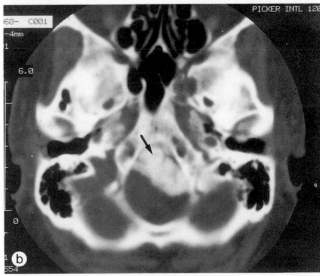

Fig. 3.16 Trigeminal neuroma. Enhanced axial (**a**), coronal (**b**) and sagittal (**c**) T1-weighted images. A large, strongly enhancing mass straddles the middle and posterior fossae. Cystic degeneration has developed in the deep aspect of the lesion (short arrows). The tumour extends through the foramen ovale into the infratemporal region (long arrow).

Fig. 3.17 Clival meningioma. (**a**) Axial CT. A densely calcified mass is seen applied to the clivus and left petrous apex, extending to overlie the porous acousticus. A lower section (**b**) shows the tumour extending caudally into the foramen magnum (arrow).

(Fig. 3.18). While hyperostosis is often present (Fig. 3.19), there may also be signs of local destruction of the bone. MR imaging shows a lesion of slight hypo- or isointensity on T1WI and of variable but often only mild hyperintensity on T2WI.

Intracranial mengingiomas may extend via basal foramina and fissures into the infratemporal and pterygopalatine fossae. Contrast-enhanced MR imaging may show a characteristic tail of enhancement within the dura adjacent to the tumour.

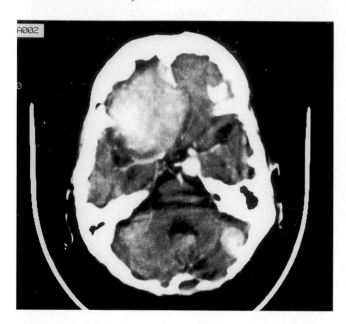

Fig. 3.18 Multiple basal meningiomas. Axial CT scan demonstrating strongly enhancing masses based upon the anterior skull base floor, left tentorial edge and occipital bone. Note previous left-sided posterior fossa surgery.

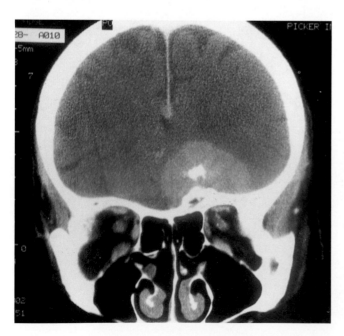

Sella tumours

Pituitary adenomas may behave in a locally aggressive manner with inferior extension into the skull base (Fig. 3.20). Although vascular lesions, enhancement may be difficult to appreciate within tumour extending inferiorly into the skull base. The well-known beam hardening artefacts in the skull-base region, prominent enhancement of normal extracranial soft tissues or the presence of tumour necrosis

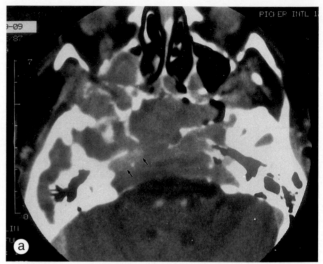

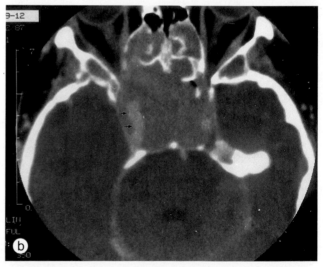

Fig. 3.20 Invasive pituitary adenoma. Axial CT scans (**a**), (**b**) show extensive destruction of skull base with involvement of the ethmoid and sphenoid sinuses, petrous apices and invasion of the cavernous sinuses. Note encasement of the right internal carotid artery (arrows).

Fig. 3.19 Coronal CT scan. Hyperostosis of the anterior cranial fossa floor is present, associated with a moderately enhancing meningioma. Note amorphous calcification within the tumour.

may all lead to problems in assessing these lesions. While CT scanning is more sensitive to bone destruction, MR imaging provides a fuller assessment of tumour involvement. Loss of the normal clival marrow fat hyperintensity on T1WI provides subtle information indicating early clival involvement.

Craniopharyngiomas

These epithelium-derived lesions arise from cell nests of Rathke's cleft and are essentially tumours of the suprasellar region, presenting predominantly in the first and second decades. In approximately 20% of cases there is sellar extension which, in a small proportion of cases, includes erosion of the sellar floor and invasion of the sphenoid sinus (Fig. 3.21).

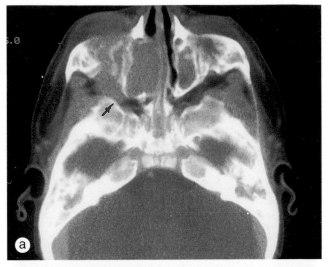

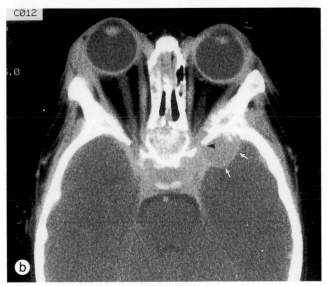

Fig. 3.22 Neuroblastoma metastases in a 2-year-old child. Axial CT scan through the level of the maxillary sinuses (**a**) and orbits (**b**). Note the right-sided deposit based upon the maxillary sinus and nasal cavity with associated bone destruction. Early involvement of the pterygopalatine fossa is present (arrow). Further deposits are seen in the right outer canthal region and left middle cranial fossa (arrows) anteriorly, with involvement of the superior orbital fissure (arrowhead).

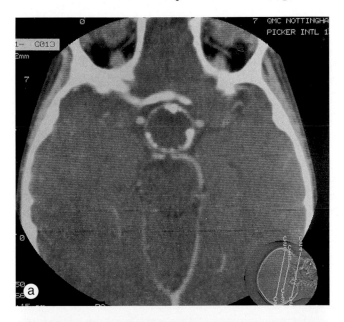

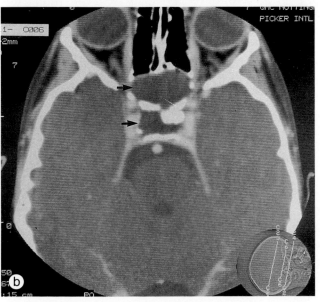

Fig. 3.21 Craniopharyngioma. Axial CT scans demonstrate a characteristic suprasellar cystic lesion with mural calcification (**a**). A lower section (**b**) shows the lesion extending inferiorly into the sella and sphenoid sinus (arrows).

Metastases and intrinsic tumours

Haematogenous spread of metastases to the dura or medial temporal lobes may lead to a non-specific mass lesion producing destruction of the skull base (Fig. 3.22). Cerebrospinal fluid seeding may produce a similar appearance but in association with evidence of leptomeningeal involvement elsewhere in the neuraxis, best demonstrated by contrast-enhanced MR imaging. Intrinsic tumours may also rarely grow exophytically to invade skull-base structures such as the cavernous sinus and Meckel's cave [15].

Osseous skull base tumours

Esthesioneuroblastoma (olfactory neuroblastoma)

These uncommon tumours arise in the region of the cribriform plate, probably originating from one of the olfactory nerves or olfactory mucosa [16]. The tumour usually grows downwards into the nasal cavity and adjacent ethmoid sinuses. In 15–20% of cases there is a significant intracranial extension [16]. Coronal CT scanning demonstrates an enhancing mass and enables assessment of the degree of intracranial extension (Fig. 3.23). Although not always apparent this should always be assumed to be present. Surgery is aimed at excising the region of the cribriform plate [17]. Disease within the ethmoid sinuses, particularly carcinoma of the sinuses and inflammatory disease with associated mucocele formation, may closely resemble these tumours. MR demonstrates non-specific hyperintensity on T2WI, but the relationship of the lesion to the cribriform plate may indicate the diagnosis.

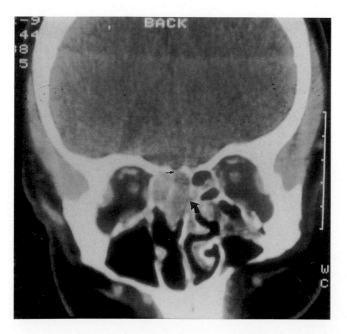

Chordomas

Chordomas are rare neoplasms arising in the vestigial remnants of the embryonic notochord. The commonest site of presentation is the sacrococcygeal region (50%), followed by the spheno-occipital region (35%) and upper cervical spine (15%) [18]. They arise most commonly in the region of the clivus centred upon the site of the spheno-occipital synchondrosis. The clivus is, however, spared in 40–50% of cases with predominantly intracranial disease, the tumour arising in notochordal remnants adjacent to the clivus such as the petroclival and parasellar regions or in the upper cervical spine [19]. Although often malignant, these lesions are slow-growing and locally invasive, not infrequently presenting in middle life with the effects of pressure on cranial nerves or the brainstem. Central extension leads to presentation with a nasopharyngeal mass, present in over 90% of cases [20]. These lesions are highly vascular and frequently contain calcifications due either to sequestration of bone fragments or to dystrophic calcium deposits. There are two main types: the typical chordoma, a lobulated translucent mucoid tumour, and the chondroid variety, which is a mixture of chondroid and chondromatous elements.

The combination of the differing histological types, variable intratumoural calcification and the not infrequent presence of haemorrhage within the tumours leads to a variable appearance of the lesion in both CT and MR imaging. On CT scanning the tumour may appear as a large, low-density lesion or a heterogeneously increased attenuation mass. There is almost invariably destruction of the bony skull base (Fig. 3.24). On MRI the mass appears well defined and rounded, being of reduced signal intensity on T1WI and clearly outlined by the residual hyperintensive signal from the fat within the remaining clivus. The tumours show heterogeneous hyperintense on T2WI and often intense enhancement with paramagnetic contrast (Fig. 3.25). The lesion may extend into the upper cervical spine and retropharyngeal space.

The prognosis for this lesion is generally poor, being surgically inexcisable and radio-resistant, with a mean survival of approximately 7 years.

Fig. 3.23 Esthesioneuroblastoma. Coronal CT showing a heterogeneously enhancing mass within the ethmoid sinuses (curved arrow). There is subtle destruction of the roof of the ethmoid (arrow).

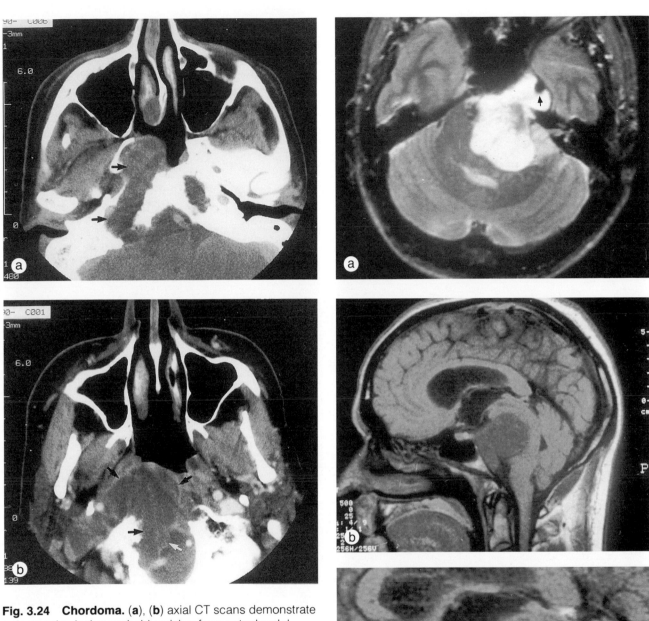

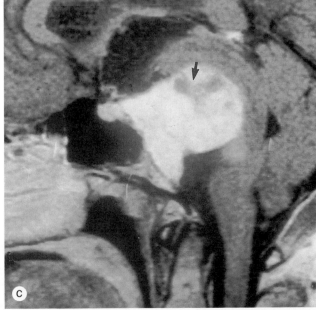

Fig. 3.24 Chordoma. (a), **(b)** axial CT scans demonstrate a destructive lesion probably arising from notochordal remnants in the petroclival region (arrows). The tumour extends medially into the sphenoid sinus and caudally into the foramen magnum and nasopharynx. Enhancement within the lesion is difficult to appreciate.

Fig. 3.25 Petroclival chordoma. (a) Axial T2-weighted image. The tumour is heterogeneously hyperintense, deeply excavating the brainstem and invading the left cavernous sinus. Note the left internal carotid artery surrounded and displaced by the tumour (arrow). **(b)** Sagittal T1-weighted image. The tumour is mildly hypointense. The brainstem compression is readily appreciated. **(c)** Intense enhancement is seen following contrast administration. Note the presence of internal non-enhancing matrix (arrow).

Chondromas and chondrosarcomas

The basiphenoid is a common site for chondromas and more rarely chondrosarcomas, centred particularly upon the spheno-occipital synchondrosis [21]. Chondrosarcomas more commonly arise within the mandible and maxilla but may also be sited in the region of sutures, e.g. the petroclival suture, and within the paranasal sinuses adjacent to the skull base (Fig. 3.26). These lesions are slow-growing and

may closely resemble chordomas. CT scanning demonstrates a locally destructive lesion with variable amounts of intratumoural calcification. MR imaging shows heterogeneous hyperintensity and, not infrequently, evidence of previous haemorrhage.

Metastases

Haematogenous spread to the skull base may take the form of localized or more widespread bone destruction, most commonly from breast, lung and renal primary tumours (Fig. 3.27). There is also

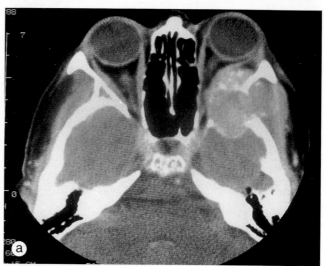

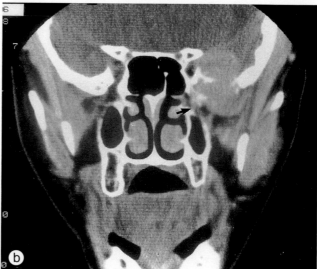

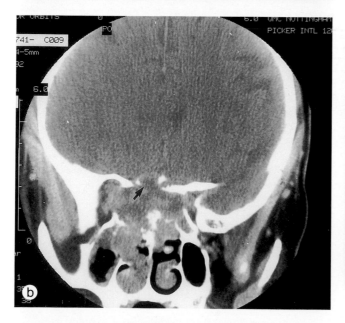

Fig. 3.26 Sphenoethmoidal chondrosarcoma. Axial (**a**) and coronal (**b**) CT scans show an expansile destructive mass containing internal stippled calcifications. There is extension into the right parasellar region (curved arrow), right orbit and destruction of the planum sphenoidale (straight arrow).

Fig. 3.27 Left sphenoid wing breast carcinoma metastasis. (**a**) Axial CT demonstrates an expansile destructive lesion bulging anteriorly into the orbit and posteriorly into the middle cranial fossa. (**b**) Coronal section demonstrating tumour encroaching into the pterygopalatine fossa (arrow) and extending superiorly into the middle cranial fossa. Lack of reaction in the overlying brain indicates that the dura remains intact.

occasional involvement by diffuse histiocytic or lymphocytic lymphoma. Prostatic and breast deposits may produce sclerosis and bony expansion with narrowing of basal foramina and compression of the contained neurovascular structures (Fig. 3.28). There is usually a soft tissue component extending into the nasopharynx or intracranial compartment. In the case of prostatic carcinoma deposits this may lead to confusion with a meningioma.

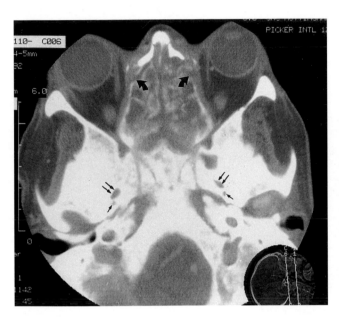

Fig. 3.28 Widespread prostatic carcinoma metastases. Axial CT scan showing diffuse sclerosis of the skull base with narrowing of the foramina spinosum (arrow) and ovale (double arrows) bilaterally. Furthermore ill-defined destructive changes are seen in the ethmoid sinuses and medial orbital walls (curved arrows) where thin sheets of tumour are present.

Extracranial tumours

Tumours arising from the nasopharynx and the paranasal sinuses may involve the skull base and are discussed in separate chapters.

Infection and inflammatory conditions

The skull base may be secondarily involved by inflammatory or infective lesions starting in the nasal cavity or paranasal sinuses. Details of these conditions are described in Chapter 5.

Vascular conditions

Aneurysms

Saccular aneurysms arising from the intracavernous portion of the internal carotid artery can enlarge to 4–5 cm in diameter with localized bone destruction, particularly involving the lesser wing of the sphenoid bone, the dorsum sellae or the petrous apex. These lesions show variable degrees of patency of the aneurysm sac, ranging from fully patent (Fig. 3.29) to largely or completely occluded sacs. Cal-

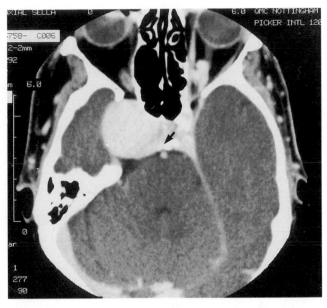

Fig. 3.29 Giant right internal carotid artery aneurysm. Axial CT showing a strongly enhancing parasellar mass with erosion of the dorsum sellae (arrow) and petrous apex.

cification of the aneurysm wall may be present. Organized thrombus within the sac often appears lamellated on scanning, but some partially occluded aneurysms may have a more homogeneous appearance which can lead to diagnostic confusion with other mass lesions, including neoplasms (Fig. 3.30). The patent lumen shows strong homogeneous enhancement following intravenous contrast administration. MR imaging demonstrates a characteristic signal void on both T1- and T2-weighted images due to turbulent or fast-flowing blood. T1-weighted images demonstrate lamellated thrombus best with alternating high- and low-signal layers [22].

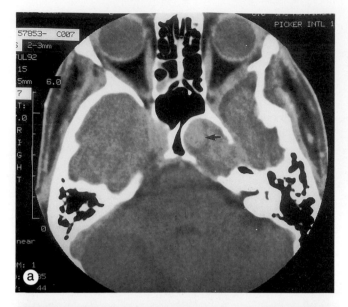

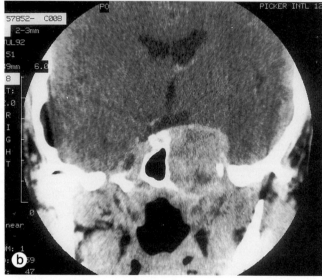

Fig. 3.30 Giant fusiform left internal carotid artery aneurysm. Axial (**a**) and coronal (**b**) CT scans demonstrate an expanding mass lesion destroying the petrous apex and bulging into the sphenoid sinus. Note the enhancing patent lumen (arrow). The bulk of the aneurysm is thrombosed, with a heterogeneous appearance.

Cavernous sinus lesions

Cavernous sinus thrombosis is usually the result of haematogenous spread of emboli from veins draining infected paranasal sinuses or periorbital tissues. CT scanning may show asymmetric expansion of the cavernous sinus and enlargement of intraorbital veins. MR imaging may show high signal intensity

on T1-weighted images from the affected cavernous sinus and ophthalmic veins. Caroticocavernous fistulae may also produce bulging of the cavernous sinus and enlarged intraorbital vessels. These lesions are most often traumatic or related to atherosclerotic degeneration and aneurysmal rupture of the intracavernous internal carotid artery. The clinical presentation is often diagnostic. CT scanning may demonstrate the extent of bony injury in traumatic cases or the presence of a large/giant aneurysm in 'spontaneous' cases. Prominent flow voids may be apparent on MR imaging, but angiography is definitive for diagnosis and, in many cases, treatment.

Trauma

Injuries to the facial region and paranasal sinuses are often accompanied by involvement of the anterior skull base. For detailed discussion of this refer to Chapter 2.

PATHOLOGY – POSTERO-LATERAL SKULL BASE

Congenital

Congenital hearing loss requires detailed investigation of the petrous bones to identify and evaluate anomalies prior to planned surgery and for genetic counselling. High-resolution CT scanning in the axial and coronal planes is the optimal mode of assessment [23] (Fig. 3.31). Abnormalities of the middle ear and external auditory canal are more prevalent than those of the inner ear. Middle ear anomalies may be classified into four major groups:

1. isolated ossicular abnormalities;
2. presence of associated external auditory canal stenosis;
3. associated atresia of the external auditory canal;
4. major dysplasia of the middle ear with or without variations in the course of the facial nerve.

Otological abnormalities are strongly correlated with head and neck syndromes such as the Klippel-Feil anomaly where conductive and sensorineural hearing loss occurs in over a third of cases. Malformations of the inner ear may involve the cochlear and vestibular aqueducts, internal auditory canal;

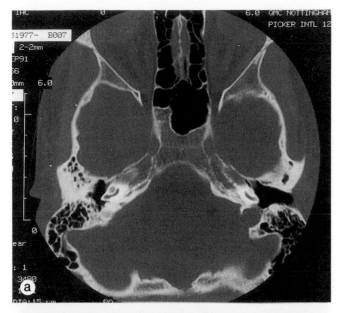

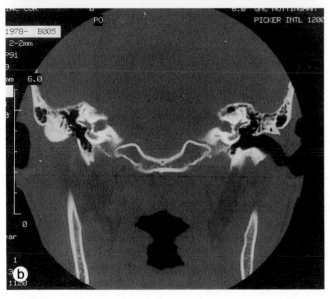

Fig. 3.31 Congenital hearing loss. Axial and coronal CT scans (**a**), (**b**) demonstrating isolated atresia of the right external auditory canal. Note preservation of the normal architecture of the labyrinthine structures.

round and oval windows and the semicircular canals, especially the lateral canal. An excellent recent study surveying congenital inner ear anomalies indicates that vestibular duct abnormalities are the most common in patients with such lesions [24].

Otodystrophic conditions involving the petrous bone include fibrous dysplasia, usually in the presence of widespread skull base involvement, and

osteogenesis imperfecta. CT scanning is optimal for the assessment of these abnormalities, which show characteristic patterns of bone density alteration and thickness.

In fibrous dysplasia the bone is homogeneously dense and thickened and there may be variable degrees of basal foraminal obliteration. Changes almost identical to otosclerosis are the hallmark of osteogenesis imperfecta, with onset of deafness in the first or second decades. Spongiotic (destructive) and sclerotic changes may be demonstrated [23].

Congenital vascular anomalies are uncommon. Venous lesions are the more prevalent lesions, and variations in the position of the jugular bulb may lead to its protrusion into the middle ear cavity where it is prone to injury during middle ear surgery. Other vascular abnormalities include an anomalous internal carotid artery and a persistent stapedial artery. An anomalous internal carotid artery courses through the middle ear where it may be recognized as a pinkish mass lying behind the tympanic membrane on direct inspection. Recognition of this entity is necessary before any exploratory surgery. The calibre of the vessel will often be reduced in comparison with the normal contralateral side.

Tumours

Neurinomas

The most commonly affected cranial nerve tumour involving the posterolateral skull base is the eighth nerve (acoustic neuroma) followed by CN VII, IX, X and XI. Acoustic neuromas are the commonest tumours affecting the temporal bone and account for 80–90% of all cerebellopontine angle tumours. Ninety per cent arise from the vestibular division and the remainder from the cochlear portion. The large majority originate within the internal auditory canal near the porous acousticus.

Large tumours are readily detected by CT scanning, the lesion being well defined and protruding into the CP cistern. These tumours usually show homogeneous enhancement. In some cases there may be internal areas of low attenuation indicating tumour degeneration or necrosis and in a small proportion of cases the tumour will be largely cystic.

Intracanalicular lesions may only be detected by appreciating focal expansion of the internal auditory canal. Until the advent of MRI, CT air cisternography was the mainstay for the demonstration of a suspected intracanalicular lesion (Fig. 3.32). More

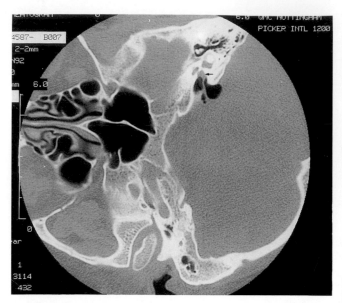

Fig. 3.32 CT-assisted air meatogram. A small intracanalicular neurinoma (arrow) is outlined by air within the internal auditory canal.

recently high definition MR imaging employing 2–3 mm slices with intravenous paramagnetic contrast has greatly increased the detection of these lesions and the more uncommon cochlear and facial nerve tumours [25]. Involvement of or growth into the internal auditory canal is best demonstrated with contrast-enhanced MRI (Fig. 3.33). CT is, however,

of particular value to the surgeon undertaking a translabyrinthine approach, providing information about the pneumatization of the mastoid air cells and the position of the facial canal, jugular bulb and lateral sinus.

Neurinomas are the commonest neoplasm of the seventh nerve, occurring most frequently in the region of the geniculate ganglion (Fig. 3.34) and descending facial canal. Neuromas of CN IX, X and XI arise from the medial aspect of the jugular foramen and tend to grow superiorly or inferiorly rather than expanding laterally.

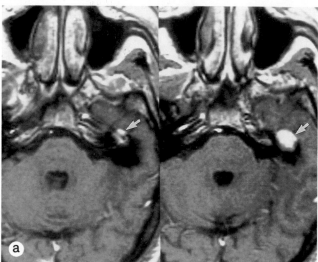

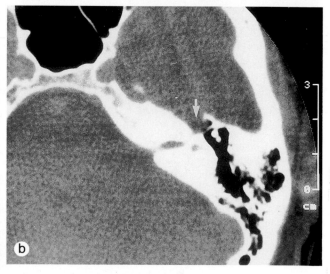

Fig. 3.34 Neuroma of the geniculate ganglion (seventh cranial nerve).. (**a**) Axial T1-weighted images after gadolinium DTPA showing a well-defined brightly enhancing mass arising from the region of the geniculate ganglion (arrow). (**b**) Axial CT section demonstrating erosion by the neuroma of the anterior petrous margin (arrow). (Reproduced by courtesy of JE Gillespie/R Fawcitt.)

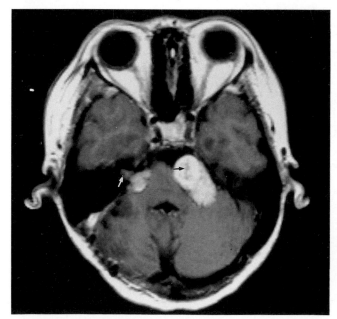

Fig. 3.33 Bilateral acoustic neurinoma. Axial enhanced T1-weighted image showing enhancing cerebellopontine angle mass lesions. On the right side tumour is seen within the internal auditory canal (white arrow). Note some internal inhomogeneity within the larger left-sided tumour in keeping with necrotic transformation (black arrow).

Paraganglionomas

This group of histochemically active tumours occurs along the distribution of glomus bodies, being most frequent in the jugular bulb or on the course of Jacobson's nerve (tympanic branch of glossopharyngeal nerve). They include glomus jugulare, glomus tympanicum, glomus vagale and carotid body tumours. The petrous bone paraganglionomas most commonly present with pulsatile tinnitus. These lesions represent the second largest group of tumours affecting the petrous bone.

Glomus tympanicum tumours arise from the mucosa of the cochlear promontory. They are usually visible otoscopically behind an intact tympanic membrane as a red or pinkish mass. CT scanning is usually sufficient to define the lesion, which appears as a soft-tissue density related to the cochlear promontory or in the floor of the middle ear cavity. There may be ossicular chain displacement or erosion of the floor of the middle ear.

Glomus jugulare tumours arise most frequently from the lateral aspect of the jugular bulb [26]. There is often widespread invasion of the skull base with involvement of the internal jugular vein and extension into the posterior fossa (Fig. 3.35). Involvement

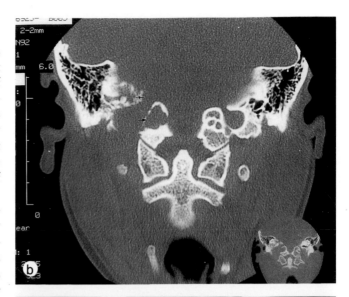

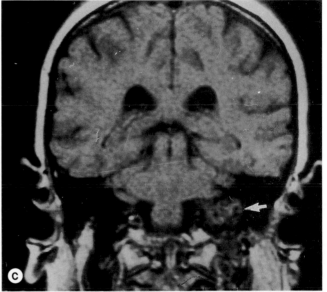

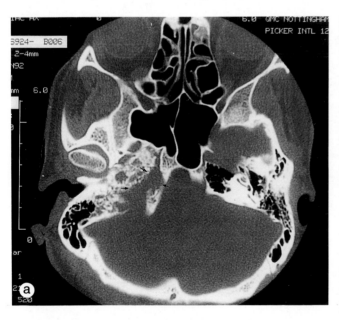

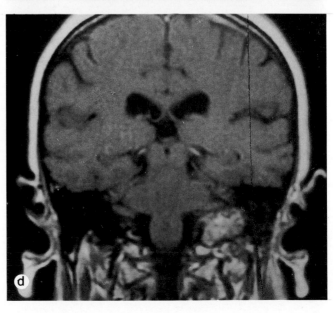

Fig. 3.35 Glomus jugulare tumour. Axial (**a**) and coronal (**b**) CT scans. A destructive process is seen based upon the jugular bulb. There is destruction of the occipital bone, infralabyrinthine segment of the petrous bone and the foramen lacerum (arrows). Coronal T1-weighted MR images pre-contrast (**c**) and post-contrast (**d**) in another patient with glomus jugulare tumour. Pre-contrast image shows a mass lesion (arrow) with signal void related to pathological blood vessels within the tumour. After contrast the tumour enhances.

of the internal carotid artery following erosion of the bony carotid canal may be seen. Angiography is required to demonstrate the integrity of the internal carotid artery, the vascular supply of the lesion (often from branches arising from both the internal and external carotid arteries) and suitability for tumour embolization or parent vessel sacrifice (Fig. 3.36). Glomus vagale tumours may extend superiorly into the skull base via the jugular foramen with associated invasion of the jugular vein and localized bone destruction.

Squamous cell carcinoma of the temporal bone

These uncommon lesions arise in the external auditory canal, middle ear cleft or mastoid cavity; however, early spread by local invasion seldom enables assessment of the origin of the tumour.

They are the commonest malignant tumour of the ear. Patients present usually in the fifth to seventh decades. Extension along the fascial planes of the Eustachian tube leads to involvement of the lateral nasopharynx while middle-ear spread or origin may be associated with superior extension to and destruction of the tegmen tympani and thence the dura and brain. MR imaging is more sensitive for assessing the extent of tumour spread, but CT demonstrates with exquisite detail the state of the external auditory canal and destructive changes within the petrous bone including the ossicular chain (Fig. 3.37). In addition, sclerosis of the middle cranial fossa floor bone may be seen as a feature of this tumour [27].

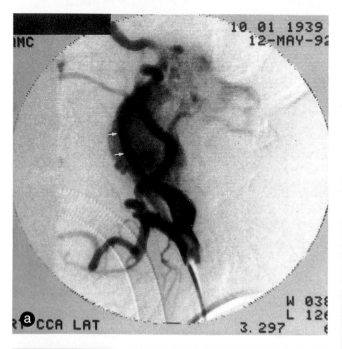

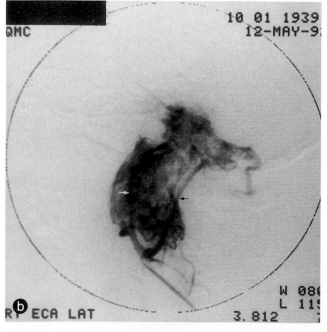

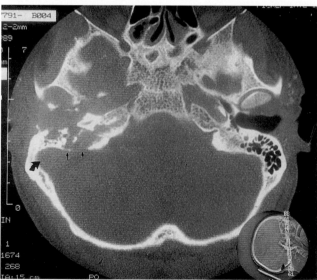

Fig. 3.37 Squamous cell carcinoma of the temporal bone. Axial CT scan demonstrating extensive bony destruction with breaching of the medial wall of the petrous bone (straight arrows) immediately adjacent to the sigmoid sinus (curved arrow).

Fig. 3.36 Glomus jugulare tumour. (a) Non-selective lateral digital subtraction angiogram of the external carotid artery showing a strong vascular blush. The internal carotid artery is encased and displaced forwards by the tumour (arrows). (b) Superselective injection demonstrating dilated ascending pharyngeal (white arrow) and posterior auricular (black arrow) arteries with intense tumour staining. Both the intratemporal and infratemporal portions of the tumour are visualized.

Cholesteatomas

There are two main types of cholesteatoma, both essentially epidermoid cysts of the petrous bone involving most frequently the middle-ear cavity and mastoid antrum.

Congenital cholesteatomas arise from ectodermal development inclusions. Acquired cholesteatomas develop from inward migration of keratinizing squamous epthelium from the external ear through the tympanic membrane into the middle ear cleft, typically via a perforation in the attic portion of the membrane. As the tumour enlarges erosion of the bony margins of the middle ear develops.

Diagnosis is usually made following otological examination. CT scanning is, however, particularly useful in assessing hidden areas such as the attic and posterior tympanic recess and in displaying anatomical variations, the position of the jugular bulge and the complications of cholesteatomas. While CT scanning is unable to differentiate granulation and other inflammatory tissue from cholesteatoma, indirect evidence will often lead to the correct diagnosis. Displacement and/or destruction of the ossicular chain and destruction of the scutum and lateral wall of the attic are principal features. Destruction of the lateral wall of the attic leads to involvement of the antrum, producing a large smooth cavity (Fig. 3.38).

Superior extension may lead to the destruction of the tegmen tympani and subsequent extradural empyema or temporal lobe abscess. Lateral spread may lead to invasion of the facial nerve canal and semicircular canals. Medial extension into the posterior fossa may result in an extradural empyema, cerebellar abscess or thrombosis within the venous sinuses.

Cystic lesions of the petrous apex are relatively rare. The most frequent aetiology is a cholesterol cyst or granuloma, lesions with a thick fibrous wall filled with altered blood and cellular debris [28]. CT scanning shows an expansile mass (Fig. 3.39) while MR characteristically demonstrates high signal on both T1- and T2-weighted images, unlike the most common cholesteatomas which are of low signal intensity on T1-weighted images.

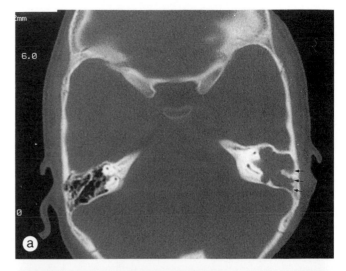

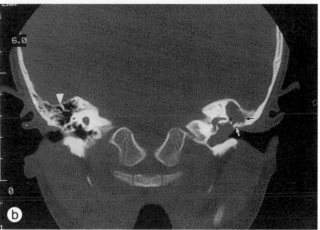

Fig. 3.38 Large left-sided acquired cholesteatoma.
Axial CT (**a**) shows an expansile, left-sided antral lesion. Note the characteristic scalloping of the mastoid bone (arrows). Coronal section (**b**) demonstrates the atticoantral mass wth destruction of the lateral wall of the attic (black arrow), ossicular chain and scutum (white arrow). Note the normal lateral attic wall on the right side (white arrowhead). Changes within the middle ear probably represent a combination of cholesteatoma and granulation tissue.

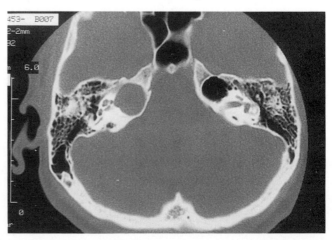

Fig. 3.39 Cholesterol granuloma. Axial CT scan demonstrating a smoothly expansile lesion within the right petrous apex. Note the extensively pneumatized ipsilateral mastoid air cells and the large contralateral petrous apex air cell.

Miscellaneous

The petrous apex may be involved by spread from nasopharyngeal carcinomas, metastases and petro-clival-based chordomas, as previously described. Meningioma confined to the intracranial portion of the posterior fossa is a well-defined entity. These may rarely originate within the temporal bone, arising from arachnoid cell nests in the internal auditory canal or jugular foramen. They may spread to involve the adjacent skull base, intracranial cavity or inferiorly into the soft tissues of the neck and the spinal canal [29]. Mucoceles of the petrous apex are similarly rare lesions which may simulate cholesterol granulomas and cholesteatomas in being erosive expansile masses without evidence of contrast enhancement [30]. Osteomas and osteochondromas may occasionally arise from the bony external auditory canal. They are best assessed by CT scanning (Fig. 3.40).

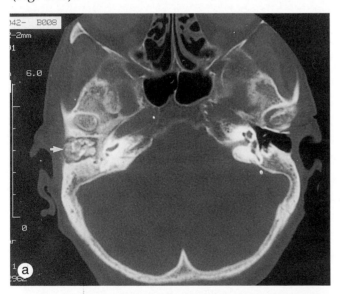

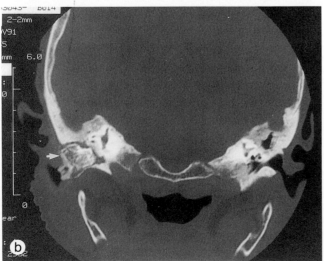

Infection

Acute otitis media is in most cases a self-limiting process. In 2–5% of cases acute coalescent or masked mastoiditis develops. If inadequately treated serious intracranial complications may follow. CT scanning in the acute stages of the disease demonstrates opacification of the mastoid air cells and, if unchecked, demineralization of the bony trabeculae. Breakdown of trabeculae and coalescence results in an irregular cavity and empyema formation. Perforation of the mastoid cortex leads to a subperiosteal abscess. The most common cause of extension of middle ear disease in non-acute cases is the presence of a cholesteatoma expanding and eroding the adjacent bony structures. Intracranial spread occurs along lines of least resistance. Typically, there is erosion of the tegmen tympani which may result in a middle fossa extradural abscess or temporal lobe abscess. Posterior fossa spread results in abscess formation in the extradural compartment and the anterior portion of the cerebellar hemisphere. Other complications of fulminant acute mastoiditis are thrombosis of the lateral, sigmoid, petrosal and cavernous sinuses, suppurative labrynthitis, petrous apicitis and facial nerve palsy [31].

'Malignant' otitis externa, a particularly virulent infection arising from the external auditory canal, occurs most often in diabetics. The pathogen is usually *Pseudomonas aeruginosa*. A necrotizing soft tissue mass with extensive bony destruction is the hallmark of this lesion (Fig. 3.41). Differentiation from a carcinoma may be difficult, but a distinct mass tends to be more a feature of the latter. Biopsy is, however, essential for definitive diagnosis.

If left unchecked, spread to the middle and inner ear, periauricular soft tissues and subcranially into the nasopharynx or oropharyanx may ensue. Intracranial spread along the cranial nerves may also occur [32]. MRI shows the extent of involvement to be greater than that apparent on CT scanning [33].

Fig. 3.40 Osteoma of the external auditory canal. Axial (**a**) and coronal (**b**) CT scans showing a pedunculated bony mass (arrows) enlarging and almost completely occluding the external auditory canal.

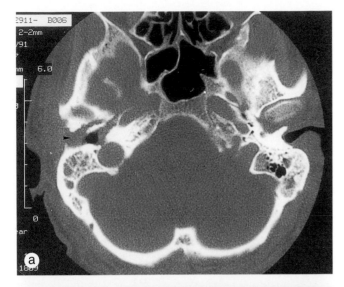

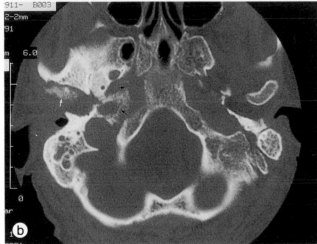

Fig. 3.41 Malignant otitis externa in an elderly male diabetic. (a) CT scan demonstrating a soft tissue mass within the external auditory canal (arrowhead) with irregularity of the bony margins and thinning of the lateral wall of the jugular bulb. **(b)** More caudal section showing destruction of the head of the mandible (white arrow) and demineralization and destruction of the petrous bone in the region of the carotid canal and foramen lacerum (black arrows).

Vascular

Vascular abnormalities involving the postero-lateral skull base will often present clinically with pulsatile tinnitus. Dural arterio-venous fistulae are probably the commonest lesions.

Other rarer causes include fibromuscular dysplasia, petrous internal carotid artery aneurysms and stenotic atherosclerotic disease. The aberrant internal carotid artery and high jugular bulb have been discussed above.

Trauma

Injury to the temporal bone occurs in approximately 20% of patients injured in a major road traffic accident. The most serious outcome of such an injury is a facial nerve palsy which may be accompanied by hearing loss [34]. CT is the optimal imaging modality for the assessment of these injuries, employing thin slice (1–2 mm) imaging in the axial and coronal planes. A preliminary scan of the brain is required to exclude damage to the central auditory nuclei and brainstem.

There are two main types of fracture affecting the petrous temporal bone. Longitudinal fractures are the commonest (80%) and extend along the long axis of the petrous bone viewed from the axial plane.

The bony labyrinth is often spared, the fracture line usually passing from the squamous temporal bone across the roof of the external auditory canal and into the tympanic cavity and tegmen tympani (Fig. 3.42). Sensorineural hearing loss is produced most often by labyrinthine concussion.

Not infrequently the fracture extends to involve the carotid artery or jugular canals, foramen lacerum or eustachian canal. The ossicular chain may be disrupted, the tympanic membrane ruptured or the middle ear cleft filled with blood, all leading to a conductive hearing loss. Facial nerve injury is less common.

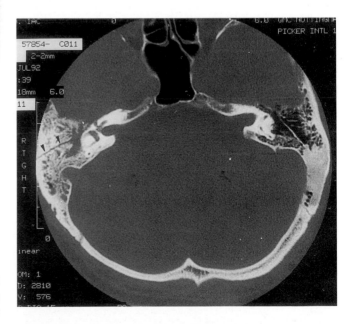

Fig. 3.42 CT scan of a longitudinal fracture of the petrous temporal bone. The fracture (arrowheads) extends along the long axis of the petrous bone into the tympanic cavity but spares the bony labyrinth. Note the diffuse opacification of the mastoid air cells and the middle ear.

Transverse fractures account for 10–15% of injuries. The axis of the fracture is perpendicular to the long axis of the petrous bone. The fracture may extend to involve the internal auditory canal, jugular foramen or bony labyrinth. Total hearing loss is found in the majority of patients suffering a transverse fracture in which there is involvement of the labyrinthine structures. Extension into the internal auditory canal, region of the geniculate ganglion or labryinthine segment of the facial nerve canal often results in damage to the proximal facial nerve (Fig. 3.43).

Complex comminuted fractures occur in 5–10% of cases, usually associated with a major head injury. Severe cerebrospinal fluid otorrhoea or herniation of the brain into the petrous bone may be seen.

REFERENCES

1. Pinto RS, George AE, Koslow M et al. Neuroradiology of basal anterior fossa (transethmoidal) encephalocoeles. Radiology 1975; 117: 78–85.
2. Diebler C, Dulac O. Cephalocoeles: clinical and neuroradiological appearance. Neuroradiology 1983; 25: 199–216.
3. Yakota A, Matsukado Y, Fuwa I, Moroki K, Nagahiro S. Anterior basal encephalocoele of the neonatal and infantile period. Neurosurgery 1986; 19: 468–478.
4. Starkman SP, Brown TC, Linell SA. Cerebral arachnoid cysts. J Neuropathol Exp Neurol 1958; 17: 484–500.
5. Frodel JL, Larrabee WF, Raisis J. The nasal dermoid. Otolaryngology – Head and Neck Surgery 1989; 101: 392–396.
6. Sessions R. Nasal dermal sinuses – new concepts and explanations. Laryngoscope 1982; 92 (Suppl 29): 1–28.
7. Fornadley JA, Tami TA. The use of magnetic resonance imaging in the diagnosis of the nasal dermal sinus–cyst. Otolaryngology – Head and Neck Surgery 1989; 101: 397–398.
8. Horowitz BL, Chari MV, James R, Bryan RN. MR of intracranial epidermoid tumours: correlation of in vivo imaging with in vitro ^{13}C spectroscopy. AJNR 1990; 11: 299–302.
9. Chiari H. Über Veränderungen des Kleinhirns infolge von Hydrocäphalie des Grosshirns. Dtsch Med Wochenschr 1891; 17: 1172–1175.
10. Naidich TP, McLone DG, Fulling KH. The Chiari II malformation: Part IV. The hindbrain deformity. Neuroradiology 1983; 25: 179–197.
11. Naidich TP, Pudlowski RM, Naidich JB. Computed tomographic signs of the Chiari II malformation: Part III. Ventricles and cisterns. Radiology 1980; 134: 657–663.
12. Wolpert SM, Anderson M, Scott RM, Kwan ESK, Runge VM. The Chiari II malformation: MR imaging evaluation. AJNR 1987; 8: 783–791.
13. Castillo M, Quencer RM, Dominguez R. Chiari III malformation: imaging features. AJNR 1992; 13: 107–113.
14. Hedemann LS, Lewinsky B, Lochridge GK et al. Primary malignant schwannoma of the gasserian ganglion. J Neurosurg 1978; 48: 279–283.

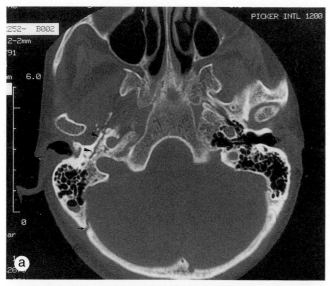
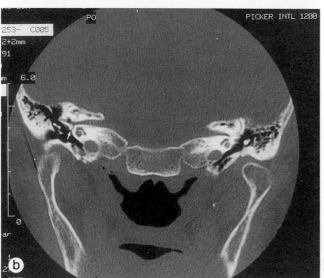

Fig. 3.43 Transverse fracture of the petrous temporal bone. (**a**) Axial CT scan. The fracture line (arrowheads) is perpendicular to the long axis of the petrous bone with sutural diastasis (arrow). (**b**) Coronal CT scan showing the fracture extending into the internal auditory canal (arrow) with involvement of the first part of the facial nerve.

15. Yuh WTC, Nguyen HD, Mayr NA, Follett KA. Pontine glioma extending to the ipsilateral cavernous sinus and Meckel's cave: MR appearance. AJNR 1992; 13: 346–348.

16. Burke DP, Gabrielsen TO, Knake JE, Seeger JF, Oberman HA. Radiology of olfactory neuroblastoma. Radiology 1980; 137: 367–372.

17. Mills SE, Frierson HF. Olfactory neuroblastoma: a clinicopathological study of 21 cases. Am J Surg Pathol 1985; 9: 317–327.

18. Batsakis JG, Kittleson AC. Chordomas. Otorhinolaryngologic presentation and diagnosis. Arch Otolaryngol 1963; 78: 168–175.

19. Sze G, Uichanco LS, Brant-Zawadzki MN, David RL, Gutin PH, Wilson CB, Norman D, Newon TH. Chordomas: MR imaging. Radiology 1988; 166: 187–191.

20. Richter HJ, Batskis JG, Boles R. Chordomas: nasopharyngeal presentation and atypical long survival. Ann Otol 1975; 84: 327–332.

21. Bahr AL, Gayler BLWL. Cranial chondrosarcoma. Radiology 1977; 12: 151–156.

22. Biondi A, Scialfa G, Scotti G. Intracranial aneurysms: MR imaging. Neuroradiology 1988; 30: 214–218.

23. Hasso AN, Ledington JA. Imaging modalities for the study of the temporal bone. Otolaryngol Clin N Amer 1988; 21: 219–245.

24. Mafee MF, Charletta D, Kumar A, Belmont H. Large vestibular aqueduct and congenital sensorineural hearing loss. AJNR 1992; 13: 805–819.

25. Brogan M, Chakeres. Gd-DTPA-enhanced MR imaging of cochlear schwannoma. AJNR 1990; 11: 407–408.

26. Lloyd GAS, Phelps PD. The investigation of petromastoid tumours by high resolution CT. Br J Radiol 1982; 55: 483–491.

27. Friedmann DP, Rao VM. MR and CT of squamous cell carcinoma of the middle ear and mastoid complex. AJNR 1991; 12: 872–874.

28. Gherini SG, Brackmann DE, Lo WWM, Solti-Bohman LG. Cholesterol granuloma of the petrous apex. Laryngoscope 1985; 95: 659–664.

29. Biggs MT, Fagan PA, Sheehy JPR, Bentivoglio PJ, Doust BD, Tonkin J. Meningioma of the posterior skull base. Skull Base Surgery 199; 1: 43–50.

30. DeLoshier HL, Parkins CW, Garek RR. Mucocele of the petrous temporal bone. J Laryngol Otol 1979; 93: 177–180.

31. Mafee MF, Singleton EL, Valvassori GE et al. Acute otomastoiditis and its complications: role of CT. Radiology 1985; 54: 391–397.

32. Mendez G, Quencer RM, Post JP et al. Malignant external otitis: a radiographic-clinical correlation. AJR 1979; 132: 957–961.

33. Gherini SG, Brackman DE, Bradley WG. Magnetic resonance imaging and computerised tomography in malignant external otitis. Laryngoscope 1986; 96: 542–548.

34. Hasso AN, Ledington JA. Traumatic injuries of the temporal bone. Otolaryngol Clin N Amer 1988; 21: 295–316.

4 Orbital imaging

A. Jackson and R. A. Fawcitt

ORBITAL ANATOMY

Muscles and connective tissues

The orbital cavity is lined by periosteum which is firmly attached to bone only at the anterior orbital rim, the orbital fissures and the margins of the optic canal. Periosteum is continuous with the meningeal sheath which surrounds the optic nerve and thickens to form a circular ligament (the annulus of Zinn) around the optic canal and the confluence of the superior and inferior orbital fissures (Fig. 4.1). The

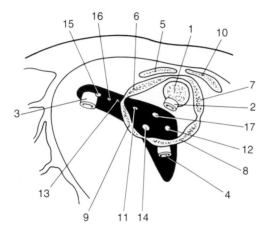

Fig. 4.1 Frontal view of the right orbital apex. The anulus of Zinn encircles the optic canal and the medial portion of the superior orbital fissure.
1 = optic nerve; 2 = ophthalmic artery; 3 = superior ophthalmic vein; 4 = inferior ophthalmic vein; 5 = levator palpebrae superioris; 6 = superior rectus; 7 = medial rectus; 8 = inferior rectus; 9 = lateral rectus; 10 = superior oblique; 11 = superior division of third nerve; 12 = inferior division of third nerve; 13 = trochlear nerve; 14 = sixth nerve; 15 = lacrimal nerve; 16 = frontal nerve; 17 = nasociliary nerve.

four rectus muscles arise from the ligament of Zinn and pass forward to insert into the scleral layer of the globe anterior to its equator. The superior and inferior oblique muscles insert into the globe posterior to its equator. The superior oblique arises from the periosteum of the medial orbital roof near the apex and passes forward to the superomedial aspect of the orbit where it loops around a cartilaginous pulley (the trochlea) before turning posterolaterally to its insertion on the globe. The inferior oblique

arises just lateral to the nasolacrimal fossa, passing posterolaterally below the globe to its insertion. Levator palpebrae superioris arises from the tendon of the superior rectus and passes forwards above it to insert into the tarsal plate of the upper lid. These two muscles cannot be distinguished even on coronal images and are usually referred to as the superior muscle group. The extraocular muscles, optic nerve sheath and orbital periosteum are interconnected by fibrovascular connective tissue septa.

The four rectus muscles and their interconnecting septa divide the orbit into the conal space lying behind the globe and the extraconal space which lies between the rectus muscles and the orbital walls. The anterior boundary of the extraconal space is formed by the orbital septum, a thick fibrous disc which arises from the orbital margins and inserts into the globe 6–7 mm behind the limbus. The septum is continuous with the scleral insertions of the rectus muscles and the inner fibrous layers of the eyelids and forms a considerable barrier to many disease processes.

Nerves

The optic nerve passes forwards from the chiasm into the optic canal where it lies superior and medial to the ophthalmic artery. It is surrounded by an anterior perineural continuation of the subarachnoid space which extends to the posterior aspect of the globe. The oculomotor, trochlear, abducens and ophthalmic (first division of the trigeminal) nerves all lie within the lateral wall of the cavernous sinus within the middle cranial fossa. The oculomotor, abducens and nasociliary division of the ophthalmic nerves all enter the orbit within the ligament of Zinn while the trochlear nerve and the lacrimal and frontal branches of the ophthalmic nerve enter the orbit outside the annulus. The trochlear nerve supplies the superior oblique and the abducens nerve supplies the lateral rectus. The oculomotor nerve supplies all the other extraocular muscles and carries autonomic fibres which innervate the iris, ciliary body and lacrimal gland.

Blood vessels

The main arterial supply to the orbit is from branches of the ophthalmic artery which passes forward above the optic nerve to lie medially. There are potential collateral connections with the external carotid artery via the lacrimal and middle meningeal arteries. Orbital veins are valveless and communicate anteriorly with the facial vein. The superior ophthalmic vein is the largest and is divided into three parts. The first part lies extraconally between the medial and superior recti, the second passes intraconally beneath the superior rectus and the third lies along the lateral border of the superior rectus passing into the superior orbital fissure. The smaller inferior ophthalmic vein passes back lateral to the inferior rectus to join the superior vein in the orbital apex.

The globe

The globe consists of three layers. The outer layer is the fibrous sclera which is continuous anteriorly with the transparent cornea. Sclera is continuous posteriorly with the dura of the optic nerve and is pierced several millimetres from the optic nerve by the ciliary vessels and nerves. The second layer of the globe, lying immediately beneath the sclera, is the uveal tract which is composed of the ciliary body and iris anteriorly and the choroid, a thick vascular layer which extends posteriorly from the ciliary body to the optic nerve head. The third and innermost layer is the retina which is composed of an outer retinal pigment epithelium and an inner sensory retinal layer.

Normal sectional anatomy as seen on MR is displayed in Figs 4.2–4.4.

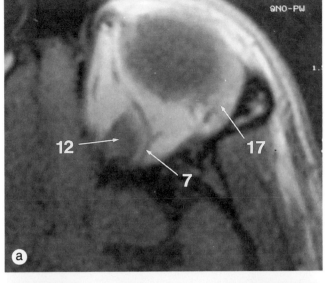

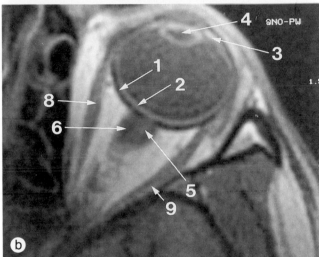

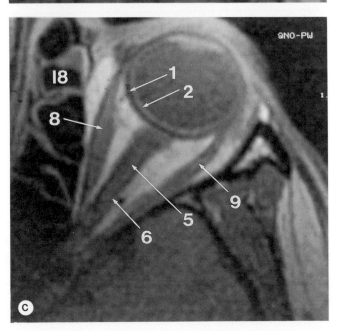

Fig. 4.2 T1-weighted transverse MRI images of the normal orbit superior (**a**) to inferior (**c**).
1 = sclera; 2 = choroid; 3 = ciliary body; 4 = lens;
5 = optic nerve; 6 = perineural space; 7 = superior ophthalmic vein; 8 = medial rectus muscle; 9 = lateral rectus muscle; 10 = superior rectus muscle; 11 = inferior rectus muscle; 12 = superior muscle complex;
13 = inferior oblique muscle; 14 = superior oblique muscle; 15 = trochlea; 16 = ophthalmic artery;
17 = lacrimal gland; 18 = ethmoid air cells; 19 = maxillary antrum; 20 = sphenoid sinus; 21 = ophthalmic artery.

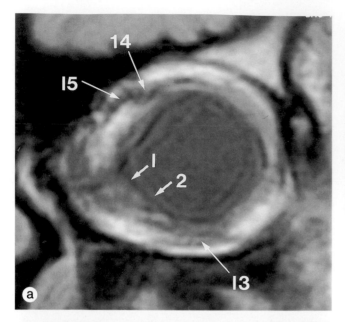

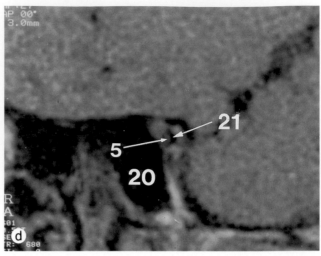

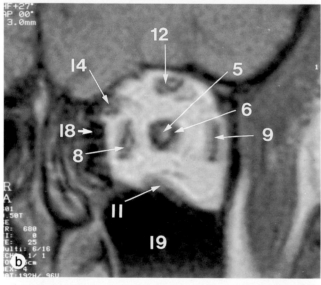

Fig. 4.3 T1-weighted coronal images of the normal orbit anterior (**a**) to posterior (**d**). Abbreviations as for Fig. 4.2.

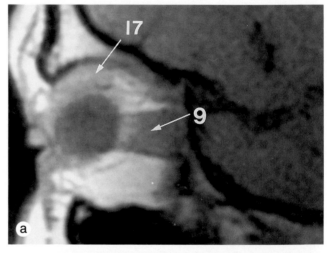

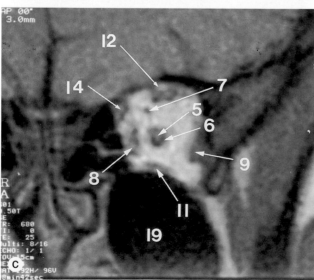

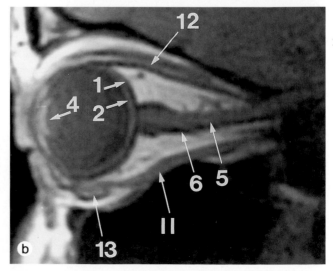

Fig. 4.4 T1-weighted oblique parasagittal images of the normal orbit lateral (**a**) to medial (**b**). Abbreviations as for Fig. 4.2.

ORBITAL IMAGING TECHNIQUES

Computed tomography (CT), magnetic resonance imaging (MRI) and orbital ultrasound form the mainstay of orbital imaging techniques [1,2,3]. Plain film radiography, orbital venography and angiography may all help to resolve specific diagnostic dilemmas but are seldom necessary in routine orbital investigations. A review of orbital ultrasound is beyond the scope of this chapter, however, ultrasound can provide highly specific diagnostic information in certain pathological states (i.e. cavernous haemangioma) and a good-quality ultrasound machine with a 5–7 MHz sector probe can provide a satisfactory alternative to a dedicated orbital unit [4].

CT imaging protocols should be designed to avoid unnecessary irradiation of the lens. Modification of exposure factors is possible on most modern machines and the presence of fat, which acts as a natural contrast agent in the orbit, enables diagnostic images to be performed satisfactorily while incurring a significantly lower radiation dose than is required for imaging of cranial soft tissues. Exact exposure factors vary depending upon the machine in use, but we find that images performed at 120 kVp and 140 mAs provide excellent image quality with a radiation load which is less than 30% of that incurred during routine cerebral imaging. The routine use of 1.5–3.0 mm thick transverse images with biplanar reformation will also remove the need for direct coronal and sagittal images in almost all cases.

MRI incurs no radiation cost and allows direct imaging in any plane required. The absence of signal from bone makes MRI particularly suited to the demonstration of the orbital apex and optic canal but renders it insensitive to pathological calcifications and bone destruction. A standard head coil is highly satisfactory for orbital imaging and provides images of the normal side for direct comparison. The use of a surface coil provides improved signal-to-noise ratio and spatial resolution and is particularly valuable in the assessment of lesions in the globe. Ocular and eyelid movements are the major causes of image degradation and the planning of imaging sequences should aim at reducing imaging time. The patient should be asked to fix his/her gaze on a single point or to keep the eyes closed. Eye make-up, particularly mascara, should be removed to avoid paramagnetic artefacts (see Fig. 4.27, p. 86). The use of fat suppression techniques can be of particular value in the orbit. Short tau inversion recovery sequences (STIR) will reduce the signal from fat and can improve the visualization of some orbital pathology but will also suppress the enhancement produced by paramagnetic contrast agents. Other fat suppression techniques, including Dixon's method (Fig. 4.5) will not suppress contrast enhancement and can considerably increase the diagnostic yield in some conditions (i.e. optic neuritis) [1].

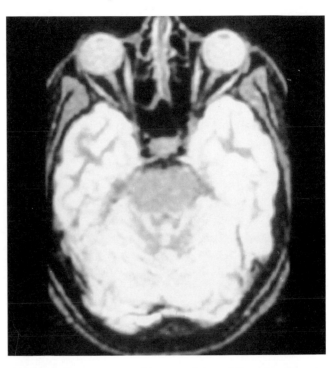

Fig. 4.5 Fat-suppressed image of the orbits using Dixon's technique.

ORBITAL PATHOLOGY

The classical division of orbital pathology into preseptal, extraconal and intraconal has some value but can be misleading in individual cases. The anatomical location of a lesion can, however, provide one of the most important clues to its nature. We have therefore retained an anatomical classification to describe disease processes which affect the globe, optic nerve, extraocular muscles and lacrimal gland and have divided other orbital pathology into mass lesions and inflammatory processes.

Lesions of the globe

Few disorders affecting the globe itself will come to the attention of the radiologist since ophthalmoscopy, retinal angiography and ocular ultrasound will often provide adequate information for clinical management. CT or MR imaging can be valuable in the investigation of ocular neoplasia and may be required for adequate assessment of the globe in the presence of choroidal or retinal detachment.

Choroidal and retinal detachment

Separation of the sensory layer of the retina from the deeper retinal pigment epithelium is known as retinal detachment. Retinal detachment may result from subretinal accumulation of fluid due to a tear in the sensory retina (rhegmatogeneous retinal detachment) or secondary to fluid accumulation from an inflammatory or neoplastic lesion (non-rhegmatogeneous retinal detachment). The commonest neoplasms producing retinal detachment in adults are malignant melanoma and choroidal haemangioma [1,3,5]. Separation of the choroid from the sclera is termed choroidal detachment and is caused by accumulation of fluid in the subchoroidal space due to inflammation, haemorrhage or transudation secondary to ocular hypotony.

It is not always possible to distinguish between retinal and choroidal detachment on CT and MR images. The sensory retina is a thin membrane beyond the resolution of CT or MRI and retinal detachment can only be recognized when there is sufficient contrast between the subretinal effusion and the adjacent vitreous. Large retinal detachments classically adopt a V shape centred on the optic disk and ending at the ciliary body (Fig. 4.6). Choroidal detachment results in separation of the entire retina and choroid and this membrane is visible on both CT and MR. Since the choroid is anchored posteriorly at the site of entry of the ciliary arteries and nerves choroidal effusions do not extend to the optic disk and appear as lentiform collections stopping short of the insertion of the optic nerve (Fig. 4.6). Very large choroidal detachments may bulge across the midline and meet above the optic nerve cup. These kissing choroidal detachments may be impossible to distinguish from retinal detachment on morphological criteria [3].

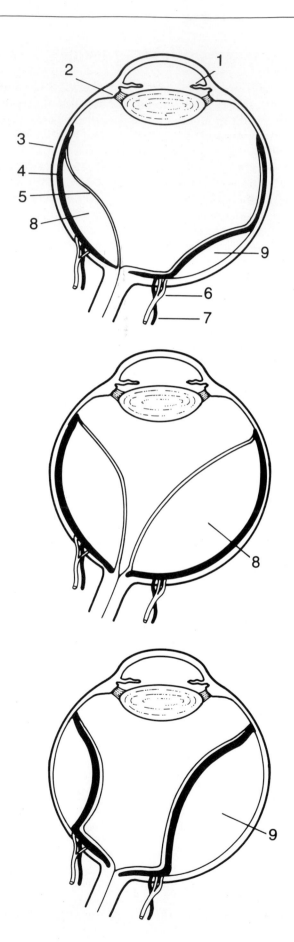

Fig. 4.6 Retinal detachment. (**a**) demonstrates the potential intraocular spaces. (**b**) The typical configuration of a large retinal detachment. Note the V-shaped appearance extending posteriorly to the optic disk (**c**) A large choroidal detachment. Note that the posterior aspect of the detachment extends medially only as far as the insertion of the ciliary arteries.

1 = iris; 2 = ciliary body; 3 = sclera; 4 = choroid;
5 = retina; 6 = ciliary arteries; 7 = ciliary nerves;
8 = subretinal effusion; 9 = subchoroidal effusion.

Haemorrhagic, or protein rich, non-rhegmatogeneous retinal detachments appear on CT as relatively dense collections in contrast to the transudates which occur in rhegmatogeneous detachment. Similar variations in the attenuation values of subchoroidal detachments allow differentiation of transudative subchoroidal effusions seen in patients with hypotony from the haemorrhagic or exudative effusions which result from inflammation or neoplasia [3,4].

Ocular melanoma

Malignant melanoma is the commonest ocular malignancy in adults and is rare in childhood. Treatment is by radical excision and enucleation, with or without orbital exenteration [3,4]. CT has been the mainstay of investigation for well over a decade and will demonstrate the tumour mass and its extent with considerable accuracy. Post-contrast images may be required to separate the tumour from a haemorrhagic retinal detachment and to satisfactorily identify any posterior tumoral extension into the optic nerve or orbit [3]. The MR appearances of pigmented melanoma are unique since the melanin within the tumour is a paramagnetic agent resulting in shortening of the T1 relaxation time (Table 4.1). These tumours therefore appear markedly hyperintense on T1-weighted images and iso to minimally hypointense on T2-weighted images [1,5] (Fig. 4.7).

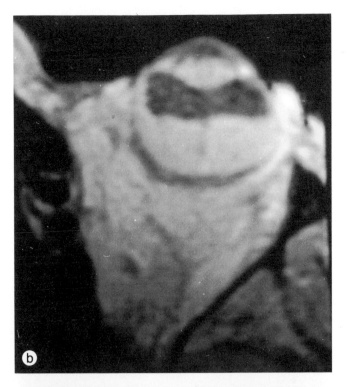

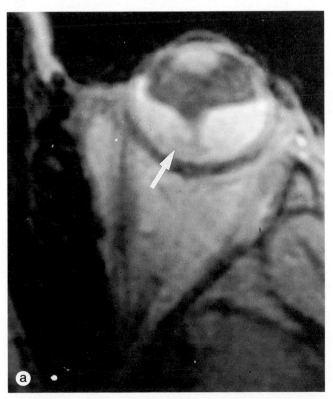

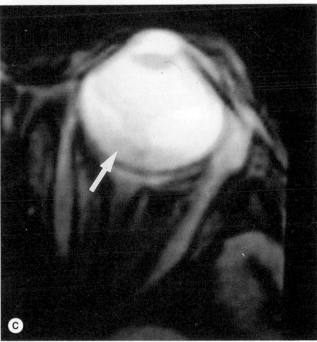

Fig. 4.7 MRI showing a non-pigmented choroidal melanoma (arrow) arising around the optic disk with a large haemorrhagic retinal detachment. (**a**) T1-weighted image. The tumour is of intermediate signal and can be distinguished from the high signal of the overlying haemorrhagic retinal detachment. (**b**) Following intravenous gadolinium the tumour enhances and becomes indistinguishable from the retinal detachment. (**c**) T2-weighted image. The tumour is isointense to the extraocular muscles.

Table 4.1   Signal intensity of ocular masses on magnetic resonance imaging

Lesion	Signal intensity relative to extraocular muscles	
	T1-weighting	T2-weighting
Ocular metastases, retinoblastoma and non-pigmented melanoma	Iso- or minimally hypointense	Iso- or slightly hyperintense
Melanotic melanoma	Hyperintense	Iso- or slightly hypointense
Choroidal haemangioma	Iso- or slightly hyperintense	Hyperintense
Early haemorrhage (intracellular methaemoglobin)	Hyperintense	Hypointense
Late haemorrhage (extracellular methaemoglobin)	Hyperintense	Hyperintense

Choroidal haemangioma

Cavernous haemangiomas of the choroid are hamartomatous lesions usually presenting in the second or third decades. They present with painless visual disturbance and may cause secondary retinal detachment due to haemorrhage. Capillary haemangiomas are more aggressive lesions histologically similar to cerebellar haemangioblastoma and 25–30% occur in patients with Von Hippel–Lindau syndrome. One-third are bilateral and multiple lesions in the same eye occur in up to 25% of cases.

CT demonstrates an enhancing mass lesion. The distinction between cavernous and capillary haemangiomas is not possible unless multiple lesions are present [2]. On MR images they may be minimally hyperintense on T1-weighted images, leading to confusion with ocular melanoma, but unlike pigmented melanoma they are usually hyperintense on T2-weighted images (Table 4.1).

Retinoblastoma

Retinoblastoma is the most common intraocular malignancy in children, accounting for 1% of childhood cancer deaths. The lesion is congenital but the average age at diagnosis is 1 year. Most cases present with strabismus or leucokoria (a pale pupillary reflex), resulting from the pale tumour mass. The tumour arises from embryonic retinal cells and demonstrates a range of histological differentiation. Release of DNA from necrotic cells leads to frequent tumoral calcification which is rare in any other ocular lesion under the age of 3 years (Fig. 4.8). Retinoblastoma may be inherited by an autosomal recessive mechanism and a spontaneous form in which bilateral tumours are associated with pineal or

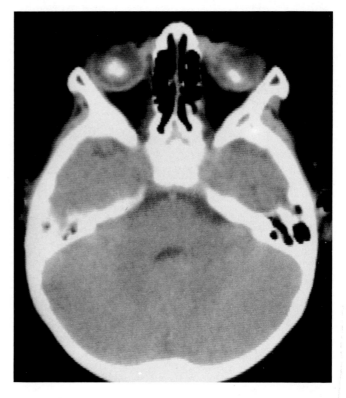

Fig. 4.8 CT of bilateral retinoblastoma showing extensive tumoral calcification.

parasellar retinoblastoma (trilateral retinoblastoma) is also recognized. Direct extension to the optic nerve or orbit and metastatic spread to bones and viscera are usually a late occurrence. Optic nerve involvement may lead to central nervous system spread and meningeal seeding. Treatment with a combination of surgery, chemotherapy and radiation results in an overall mortality of only 10–20% [1,3,5].

The tumour appears on CT as a homogeneous enhancing mass with evidence of calcification in over 90%. Most patients will have an associated retinal detachment which can usually be distinguished from the tumour on contrast-enhanced images. CT will demonstrate optic nerve and orbital involvement and images of the central nervous system should also be routinely obtained to detect ectopic tumours or metastases. MRI is superior to CT in demonstrating extraocular spread and can distinguish between tumour and retinal effusions without the use of contrast. MRI is, however, often less specific than CT since tumour calcification is poorly demonstrated.

Ocular metastases

In adult patients over 90% of orbital metastases from systemic malignancy occur in the globe. Most deposit within the vascular uveal tract and associated retinal detachment is common. The CT and MRI appearances are indistinguishable from non-pigmented ocular melanoma [1,2,3,6] (Table 4.1).

LESIONS OF THE OPTIC NERVE

Expansion of the optic nerve or nerve sheath can result from a number of neoplastic and non-neoplastic causes (Table 4.2). Neoplastic causes are rare with the exception of optic nerve glioma and perioptic meningioma. Non-neoplastic causes are commonly encountered but rarely cause diagnostic confusion since most are readily distinguished on clinical or imaging grounds [2,5].

Optic nerve glioma of childhood

Optic nerve gliomas are primary glial tumours that arise from the substance of the optic nerve or chiasm. They account for approximately 5% of orbital tumours and occur principally in childhood and adolescence with 50% presenting before the age of 5 years. Optic nerve glioma is slightly commoner in females and is associated with neurofibromatosis in up to 50% of cases [2,5] (Fig. 4.9). Optic nerve gliomas are well-differentiated, extremely slow-growing glial tumours. The tumour does not breach the dura and extension within the orbit is limited by the perineural sheath. Cystic and mucinous degeneration are common, as is hyperplasia of the arachnoid surrounding the tumour (arachnoid gliomatosis), but calcification and malignant degeneration are rare. Tumour spread occurs along the optic pathway

Table 4.2 Enlargement of the optic nerve

Primary neoplastic	Optic nerve glioma Perioptic meningioma Plexiform neurofibroma
Systemic neoplastic	Metastases Leukaemia Lymphoma
Neoplastic from globe	Retinoblastoma Melanoma
Non-neoplastic Sheath expansion	Apical crowding in thyroid orbitopathy Pseudotumour cerebri Chronic papilloedema Post-traumatic subarachnoid haematoma Arachnoid cyst
Nerve expansion	Optic neuritis Pseudotumour Infarction of optic nerve Wegener's granulomatosis Sarcoid Tuberculosis Toxoplasmosis Syphilis

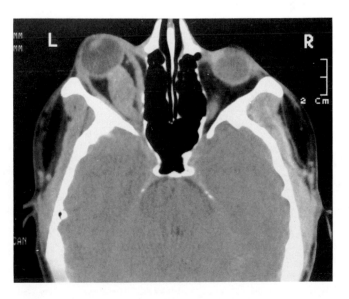

Fig. 4.9 CT showing a left optic nerve glioma in a patient with neurofibromatosis. Transverse image, showing smooth enlargement of the left optic nerve.

to involve the chiasm and optic radiations. Within the cranial cavity the tumour may also spread directly to parachiasmal structures.

Approximately 25% of optic nerve gliomas arise in, and remain confined to, the intraorbital optic nerve. Progressive but usually severe visual loss and the insidious development of proptosis are typical. Treatment is initially conservative with radical resection if there is rapid growth or disfiguring proptosis. Intracranial extension is present in approximately 10% of cases at presentation and forms an indication for prompt tumour resection if involvement of the chiasm has not occurred. Approximately 45% of optic nerve gliomas originate in the chiasm and a further 20–25% will involve it. Gliomas arising posteriorly within the chiasm carry a significantly worse prognosis and precocious puberty or hydrocephalus may result from invasion of the hypothalamus and ventricular system. Surgical excision is inappropriate and chiasmal biopsy may not only destroy remaining vision but has been associated with significant morbidity and mortality. Approximately 25% of optic nerve gliomas have multicentric origins; a third of these will present with bilateral optic nerve tumours with the remainder involving the intracranial optic pathways. Multicentric optic nerve glioma appears to be pathognomonic for neurofibromatosis and carries a particularly good prognosis for vision.

Radiological features

Radiological investigation is directed at confirming the diagnosis of optic nerve glioma and at delineating the extent of tumour with particular reference to the presence of chiasmal involvement. Optic nerve meningioma forms the major differential diagnosis and differentiating features are listed in Table 4.3. Plain films will demonstrate concentric widening of the optic canal in almost all cases (90%) where extension through the canal is present and erosion of the chiasmatic sulcus (J-shaped sella) may also be present.

CT shows a well-defined fusiform enlargement of the optic nerve with smooth margins (Fig. 4.9). Kinking and buckling of the expanded nerve is characteristic and large tumours may arise eccentrically and mimic an encapsulated intraconal tumour. Tumour attenuation and enhancement are typically heterogeneous, reflecting cystic and mucinous degeneration [2,3,5].

MRI is the preferred investigation in patients with optic nerve glioma. On T1-weighted sequences the tumour is isointense to white matter, but the signal intensities on T2-weighted sequences vary depending on the site of involvement. Tumours of the optic

Table 4.3 Comparison of optic nerve glioma and perioptic meningioma

Optic nerve glioma	Perioptic meningioma
EPIDEMIOLOGY	
Age	
Median 5 years (5% present after 20)	Median 42 years (4% present under 20)
Sex	
Slight female predominance	Over 65% in females
Associations	
20–50% have neurofibromatosis	Associated with neurofibromatosis only in children
CT FEATURES	
Optic canal enlargement common (90%)	Optic canal enlargement rare (10–12%)
Intact dura with smooth margins	Irregular margins if dura invaded
Calcification rare	Calcification common
Heterogeneous texture and enhancement due to cystic degeneration	Homogeneous texture and enhancement
Kinking of nerve	Straight, splinted nerve
	Tram track sign (50–100%)
	Bone changes at apex
	Pneumosinus dilatans
MRI FEATURES	
Nerve enlargement	Sheath enlargement
Surrounding perineural CSF space	CSF space ablated or distorted
Intradural growth	Extradural extension common
Variable enhancement	Prominent enhancement with Gd DTPA
Arachnoid gliomatosis with peripheral enhancement	
Extension along neural pathways	Extension into intraconal space and *en-plaque* on to periosteum and contralateral optic nerve

nerve are isointense to white matter on T2-weighted sequences (Fig. 4.10) but where the lesion involves the chiasm there is an abrupt increase in signal intensity. With gadolinium profound peripheral enhancement of areas of arachnoid gliomatosis may be present and should not be misinterpreted as *en plaque* meningioma or a basal meningitis [1,2].

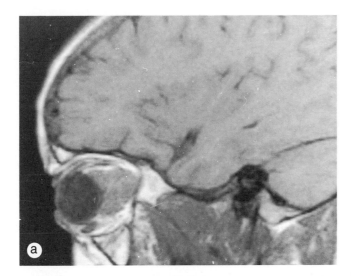

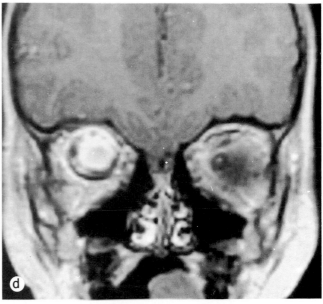

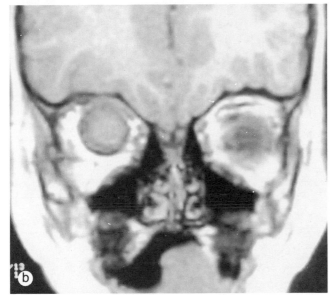

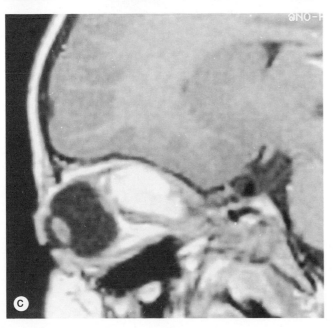

Fig. 4.10 T1-weighted sagittal (**a**) and coronal (**b**) MR images showing a large optic nerve glioma of the intraorbital optic nerve. (**c**) and (**d**) T1-weighted images showing marked tumoral enhancement following intravenous gadolinium.

Malignant optic nerve glioma of adulthood

Glioblastoma of the optic nerve represents a distinct clinical entity. Presenting in middle age, it causes rapid painful visual deterioration mimicking optic neuritis but causing binocular symptoms due to involvement of the optic chiasm. The tumour progresses rapidly, invading surrounding brain and causing focal neurological deficit. There is no effective treatment and the prognosis is uniformly fatal. CT will demonstrate enlargement of the optic nerve or chiasm in most cases but MRI appearances have not been described [2,5].

Optic nerve meningioma

Two-thirds of optic nerve meningiomas occur in women, most commonly between the ages of 30 and 60 years, although a significant proportion occur in children (4–5% under 20 years) where they show more aggressive behaviour and are usually associated with neurofibromatosis [1,2,5,6]. These tumours grow slowly and spread by local extension along the lines of least resistance. Growth may be entirely subdural, but in many cases the perineural meninges are breached and the tumour presents as an extradural, intraconal mass. Although impairment of visual acuity and proptosis are both features of optic nerve meningiomas they are invariably mild, particularly in the early stages. Progression occurs over the course of years and eventually leads to severe constriction of the visual fields with small residual islands of vision. Management is conservative in keeping with the slow rate of growth but surgical excision should be considered if there is rapid progression or if there is evidence, or risk, of intracranial spread.

Radiological features

Widening of the optic foramen or sclerotic changes in the walls of the optic canal are demonstrated in up to 12% of patients with optic nerve meningioma. CT usually shows well-defined tubular thickening (64%) or fusiform swelling (23%) of the optic nerve (Figs 4.11–4.13). In the remaining patients the

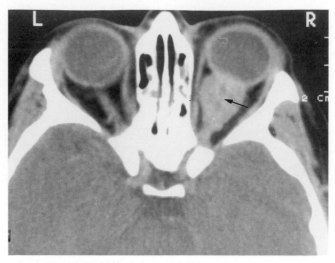

Fig. 4.12 Post-contrast CT showing a large optic meningioma. Note the central area of non-enhancement (tram track sign) corresponding to the optic nerve (arrow).

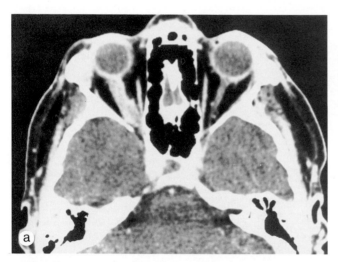

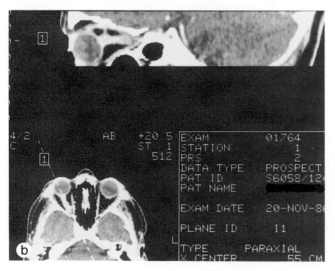

Fig. 4.11 (**a**) Post-contrast CT of right optic nerve meningioma. (**b**) Parasagittal reformation showing enlargement and enhancement of the optic nerve extending from the globe to the orbital apex.

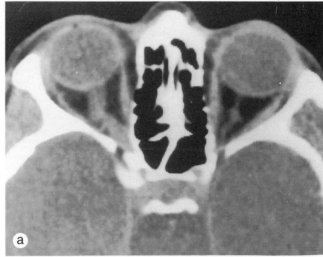

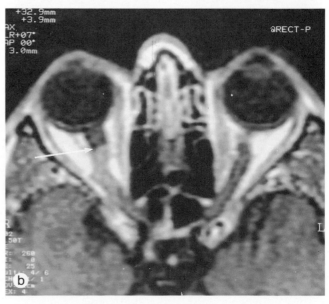

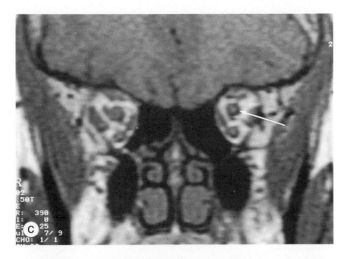

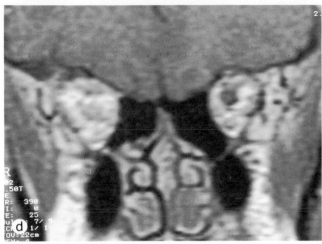

Fig. 4.13 (**a**) CT showing thickening and kinking of the right optic nerve due to an intradural optic nerve meningioma. (**b**) Transverse T1-weighted MR image following gadolinium showing the anterior extent of tumour (arrow). (**c**) Coronal T1-weighted MR showing enlargement of the right optic nerve with loss of the perineural space. The normal optic nerve and perineural space (arrow) are seen on the normal side. (**d**) Coronal T1-weighted image following gadolinium showing marked tumour enhancement.

tumour forms a prominent extradural mass in contact with the optic nerve and may be difficult to distinguish from other intraconal mass lesions. The nerve sheath complex is usually straight or splinted and does not show the kinking seen with optic nerve glioma. The margins of the optic nerve may be irregular, indicating extradural extension of tumour, and speckled or ring-shaped calcifications may be demonstrated (50%). One characteristic feature is the presence of a linear defect caused by the optic nerve passing through the mass (Fig. 4.12). This 'tram track' sign is not a specific feature of perioptic meningioma [1,7] (Table 4.4, Fig. 4.34, p. 93) but

Table 4.4 Causes of the tram track sign in optic nerve masses

Perioptic meningioma
Optic neuritis
Cavernous haemangioma
Pseudotumour
Sarcoid
Leukaemia
Lymphoma
Metastases
Perioptic haemorrhage
Normal variant

can be valuable in excluding optic nerve glioma since it does not occur in this condition. On MRI perioptic meningiomas appear isointense to muscle on T1-weighted images and isointense to fat on T2-weighted images. Intravenous paramagnetic contrast causes considerable tumoral enhancement and allows far more accurate delineation of the extent of tumour spread. MRI is particularly valuable in assessing the posterior extent of the tumour since it is not subject to the artefacts arising from bony structures which are seen on CT but fails to demonstrate areas of tumoral calcification and of cortical hyperostosis. Chemical shift misregistration artefact may also cause problems in the delineation of small perineural lesions [1,2].

Other causes of optic nerve enlargement

The majority of other optic nerve tumours are metastatic or involve the nerve by direct spread from the globe (Table 4.2). Melanoma and retinoblastoma are the most common sources of this type of optic nerve involvement and all are readily demonstrated on CT or MRI [1,5,8]. Increase in the diameter of the optic nerve sheath may result from increased cerebrospinal fluid pressure due to intracranial disease or compression in the orbital apex. Arachnoid cysts within the perioptic sheath are also well described and can cause a severe compressive optic neuropathy which is easily treated by creation of a window in the optic nerve sheath. MRI clearly demonstrates a distended cerebrospinal fluid space and is therefore the imaging modality of choice [1]. Where MRI is not available then CT, following introduction of contrast into the basal cisterns, will demonstrate the perineural space and optic nerve [2.3] (Fig. 4.14). Inflammatory pseudotumour and orbital cellulitis may lead to thickening of the optic

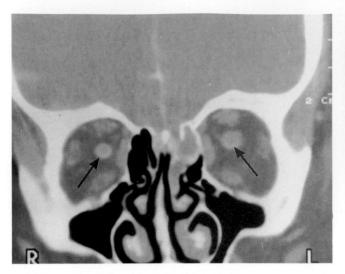

Fig. 4.14 Optic neurogram demonstrating contrast within the perineural space following intrathecal contrast injection (arrows).

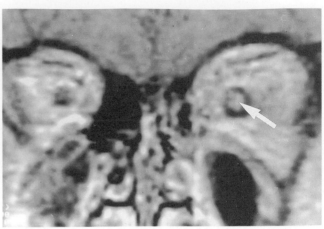

Fig. 4.15 Coronal T1-weighted MR following gadolinium administration in a patient with unilateral optic neuritis. Note the enlargement and enhancement of the left optic nerve (arrow).

nerve sheath itself [2,5,9,10]. MRI will usually demonstrate the normal optic nerve and perioptic cerebrospinal fluid space while on CT this peripheral inflammation often gives rise to the tram track sign [2,5,7]. Inflammatory expansion of the optic nerve itself is described in sarcoidosis, toxoplasmosis, tuberculosis and syphilis. These lesion closely mimic the CT appearances of perioptic meningioma and the diagnosis is usually derived from other clinical indicators of the systemic disease.

Optic neuritis

The optic nerve is a favoured site for demyelination. Clinically the patient loses visual acuity rapidly over a period of hours to days; involvement may be bilateral and may progress to total blindness. Recovery occurs in a period of weeks in most patients and is accelerated by steroid therapy. Between 40% and 50% of these patients will eventually develop multiple sclerosis and MRI of the brain will demonstrate central nervous system demyelination in 20–25% at presentation. CT demonstrates mild nerve swelling in approximately 10% of cases and post-contrast images may demonstrate central or peripheral enhancement. Conventional spin-echo MR images are usually normal but areas of high signal may be demonstrated on T2-weighted images. The use of STIR imaging increases the sensitivity of MRI to approximately 50–65%. Acute plaques will enhance following intravenous paramagnetic contrast, which also improves the sensitivity of MRI (Fig. 4.15) and the use of post-contrast fat-suppressed MR images will increase the sensitivity of MRI to approximately 90% where these sequences are available [1,2,3].

Lesions of the extraocular muscles

Enlargement of the extraocular muscles can result from direct involvement by inflammatory or neoplastic disorders or from vascular engorgement secondary to raised ocular venous pressure (Table 4.5). Thyroid orbitopathy and orbital myositis account for over 70% of cases.

Table 4.5   Causes of extraocular muscle enlargement

Inflammatory	Graves disease
	Myositic pseudotumour
	Bacterial from sinus infection
Vascular congestion	Carotid–cavernous sinus fistula
	Dural AVM
	Superior ophthalmic vein thrombosis
	Apical tumour
Neoplasm	Metastasis (especially breast)
	Lymphoma
	Leukaemia
	Rhabdomyosarcoma
Trauma	Oedema
	Haematoma
Infiltrative	Acromegaly
	Erdheim–Chesters disease

Thyroid (Graves) orbitopathy

Thyroid orbitopathy is the most common cause of proptosis and the most commonly encountered disease of the orbit in adults. It occurs in genetically predisposed individuals, is linked to the HLA-DR3 antigen and affects females four to five times more often than males. Although most commonly associated with hyperthyroidism it also occurs in hypothyroidism (10%), euthyroid Graves disease or in patients with normal thyroid function (ophthalmic Graves disease). Clinical onset of ophthalmic symptoms most commonly occurs within 18 months of the onset of hyperthyroidism but may precede thyroid abnormality [5]. The extraocular muscles are infiltrated with inflammatory cells and mucopolysaccharide deposits. In the later stages the muscle belly is replaced by fatty infiltration and focal scarring.

Mild orbitopathy is usually associated with stare and lid lag due to failure or relaxation of the lid muscles. More severe orbitopathy may cause soft tissue features, myopathy or the crowded orbital apex syndrome. Soft tissue changes include vascular engorgement, chemosis, lid oedema, epiphora and lacrimal gland enlargement. Severe orbitopathy with crowding of the orbital apex will occur in 5–10% of affected patients. Males, diabetics and older patients with a later onset of thyroid disease appear to be most at risk and present with an insidious progressive optic neuropathy. If sight is threatened then urgent medical treatment is required and surgical decompression of the orbit may be necessary [2,11–13] (Fig. 4.16).

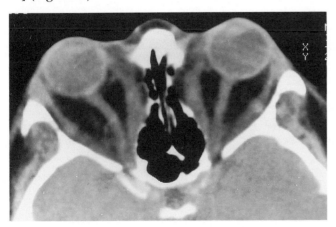

Fig. 4.16 CT following bilateral surgical decompression in a patient with thyroid orbitopathy.

Radiological features

Enlargement of extraocular muscles is demonstrated in over 90% of patients with thyroid orbitopathy and in up to 70% of asymptomatic hyperthyroid patients.

Isolated involvement of a single muscle is seen in only 5–7% and involvement is bilateral in over 80%. The inferior (63%) and medial (61%) rectus are the muscles most commonly involved. The increase in muscle bulk characteristically spares the tendinous insertions, but several workers have reported cases with tendinous thickening, making this an unreliable feature for the differentiation of other causes of muscle enlargement. In the early part of the disease the enlarged muscles may show areas of low attenuation on CT which are believed to correspond to foci of inflammatory cell infiltration. In long-standing disease areas of low attenuation corresponding to fatty infiltration may develop and can replace the entire muscle belly. In approximately 50% of patients there is also an increase in the amount of retro-orbital fat, which contributes to proptosis and is the only abnormality in 8–10% [1,2,5,13]. In patients with optic neuropathy CT is essential to confirm the presence of orbital apex crowding (Fig. 4.17). These patients demonstrate greater muscle enlargement and more significant proptosis than

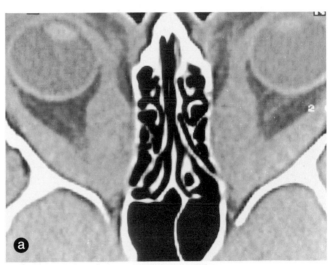

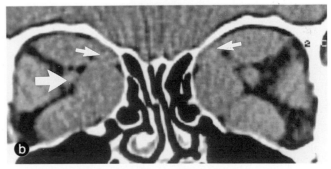

Fig. 4.17 (**a**) Transverse CT showing bilateral enlargement of the medial and lateral rectus muscles due to thyroid orbitopathy. (**b**) Coronal CT showing enlargement of the superior oblique (arrows) and rectus muscles with crowding of the optic nerve (large arrow).

patients without neuropathy [11,13]. Anterior displacement of the lacrimal gland, and enlargement of the superior ophthalmic vein and retrobulbar optic nerve are also significantly related to the presence of optic neuropathy.

MRI demonstrates morphological alterations similar to those described on CT. Involved muscles are isointense or slightly hyperintense to normal muscle on T1-weighted images and areas of high signal corresponding to fatty infiltration may be seen (Fig. 4.18). This may allow differentiation from inflammatory myositis where fatty infiltration does not occur (Table 4.6). On T2-weighted images one group has reported an increase in the measured T2 values in 50% of affected muscles, giving rise to a visible increase in signal intensity in nine of 23 patients in their study. These workers postulated that this increase in T2 corresponds to inflammation-related oedema and might predict a response to anti-inflammatory therapy [1].

Orbital Myositis

Orbital myositis forms a part of the spectrum of idiopathic orbital inflammation which is described in detail below. The disorder appears to be mediated by an autoimmune mechanism and has been reported in conjunction with collagen vascular diseases. Most cases respond dramatically to a relatively low dose of steroids but relapse is common following withdrawal of treatment [2,5,9,12]. CT demonstrates extraocular muscle enlargement which is rarely bilateral and which involves more than one muscle in 50–60% of cases (Fig. 4.19). Muscle en-

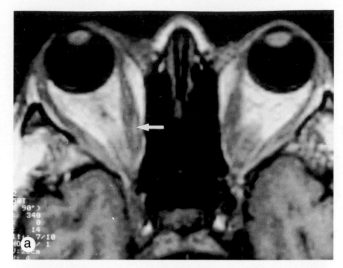
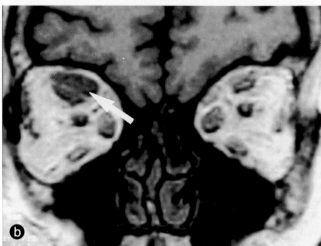

Fig. 4.18 (**a**) Axial T1-weighted MR in chronic thyroid orbitopathy showing streaks of fatty infiltration in the medial rectus muscle (arrow). (**b**) Coronal T1-weighted MR shows particular enlargement of the right levator palpebrae superioris muscle (arrow). Areas of high signal within the other extraocular muscles represent fatty infiltration.

Table 4.6   Magnetic resonance features in extraocular muscle enlargement

Lesion	T1-weighted image	T2-weighted image
Thyroid orbitopathy	Isointense to muscle Chronic cases may have areas of hyperintensity due to fatty infiltration	Isointense to muscle 20–40% have areas of high signal due to inflammatory oedema
Orbital myositis	Isointense	Isointense
Metastases (carcinoma)	Isointense	Hyperintense
Metastases (myeloma, lymphoma, leukaemia)	Isointense	Hypo- to minimally hyperintense
Vascular congestion	Isointense	Isointense
Acromegaly	Isointense	Isointense
Intramuscular haematoma	Hyperintense	Hyperintense

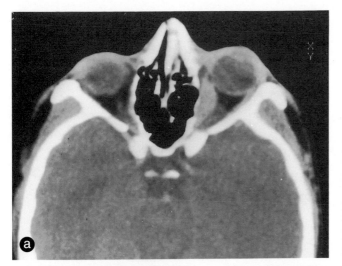

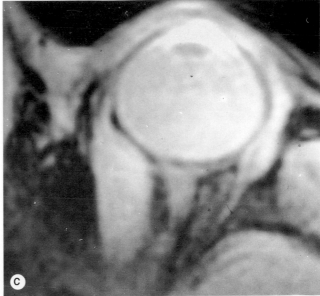

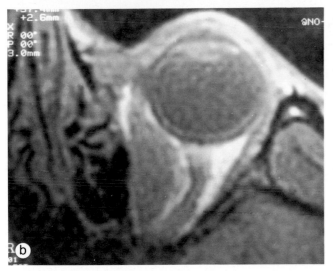

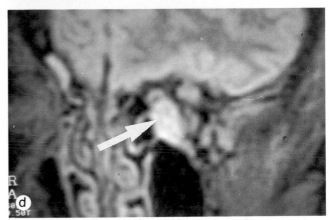

Fig. 4.19 (**a**) CT showing enlargement of the left medial rectus due to idiopathic orbital myositis. The enlarged muscle is isointense to normal extraocular muscles on both T1- (**b**) and T2- (**c**) weighted images. (**d**) STIR image demonstrating increased signal in the enlarged medial rectus (arrow).

largement usually extends to include the tendinous insertions and the margins of the muscle are often poorly defined with involvement of the retro-orbital fat in the inflammatory process (Table 4.7). Inflammatory thickening may also extend to the sclera or to the orbital apex and is particularly well seen on contrast enhanced images.

MRI signal characteristics are very similar to those seen in thyroid orbitopathy (Table 4.6, Fig. 4.19), although the areas of high signal on T1-weighted images which are seen in thyroid orbitopathy do not occur. Post-contrast images demonstrate significant enhancement within affected structures and may demonstrate otherwise occult involvement of tendinous insertions and adjacent orbital structures [1].

Other causes of extraocular muscle enlargement

Myositis resulting from direct bacterial infection can result in enlargement of extraocular muscles indistinguishable from idiopathic orbital myositis. CT usually demonstrates inflammatory changes involving adjacent structures together with sinus disease or subperiosteal abscess formation. Direct malignant invasion accounts for extraocular muscle enlargement in 5–10% of cases but is associated with evidence of bone destruction or tumour extension in most. Neoplastic lesions entirely confined to extraocular muscles are rare but have been described with metastatic carcinoma (particularly breast), rhabdomyosarcoma, lymphoma, leukaemia and multiple myeloma [2,7]. MRI may provide diagnostic information since most carcinomatous metastases have a high signal intensity on T2-weighted images (Table 4.6); however small-cell tumours such as myeloma, leukaemia and lymphoma may be indistinguishable from orbital myositis [1].

Table 4.7 Comparison of thyroid orbitopathy and orbital myositis

	Orbital myositis	Thyroid orbitopathy
Clinical features		
Onset	Rapid	Slow
Pain	Usual, related to eye movement	Rare
Eye movement	Limited and painful	Limited but painless
Response to steroids	Dramatic, complete resolution	Slow and incomplete
CT features		
Bilateral involvement	Unusual (20–50%)	Common (80%)
Single muscle	40–50%	5–7%
Muscle borders	Irregular	Smooth
Orbital fat	Commonly involved	Not involved
Tendinous insertions	Commonly involved	Involvement reported but rare
Scleral involvement	Occasional	Not involved
Site	Superior and medial most common	Inferior and medial most frequent
MRI features		
T1-weighted	Homogeneous, isointense to muscle	May contain areas of high signal due to fatty infiltration
T2-weighted	Isointense or slightly hyperintense to normal muscle	20–40% show areas of increased signal

Increased orbital venous pressure causing muscle swelling may be seen with ophthalmic vein or cavernous sinus thrombosis which usually occur secondary to sepsis. CT demonstrates generalized muscle enlargement which will show moderate post-contrast enhancement [14] (Fig. 4.20). There may be

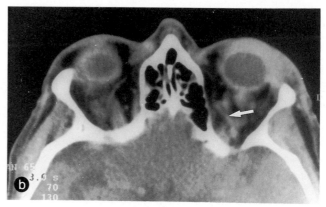

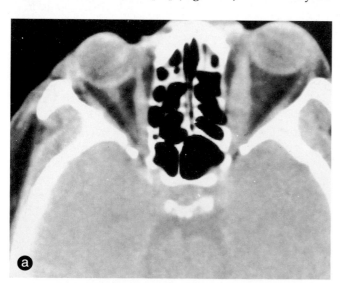

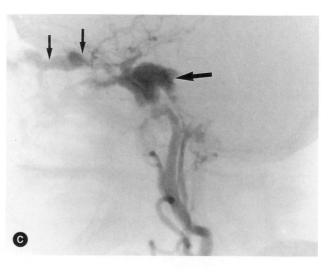

Fig. 4.20 (**a**) and (**b**) CT showing enlargement of the left medial rectus muscle and superior ophthalmic vein (arrow). (**c**) Left carotid angiogram showing a carotico-cavernous sinus fistula with early filling of the cavernous sinus (large arrow) and retrograde flow through the distended superior ophthalmic vein (small arrows).

evidence of venous congestion, with increase in size of the superior ophthalmic vein, and post-contrast images may show thrombosis as filling defects within the vein or cavernous sinus. Carotid cavernous sinus fistulas and dural arteriovenous malformations will also cause orbital venous engorgement and enlargement of extraocular muscles. The site of the shunt is almost always within the cranial cavity with the ophthalmic veins draining the shunt by retrograde flow. CT findings are non-specific and carotid angiography is required to define the site of the shunt and to plan treatment. Endovascular occlusion of carotid cavernous sinus fistulas by balloon or coil occlusion carries a minimal risk of morbidity with sparing of the parent artery and is now the treatment of choice when technically feasible [1,2,3,5].

Lesions of the subperiosteal space

Lesions arising below the periosteum are a common cause of orbital abnormality. Inflammatory and neoplastic lesions of the sinuses may involve the orbit by direct invasion. Frontal and ethmoidal mucoceles also commonly cause distortion of the orbital outline and visual disturbances (Chapter 5). Benign bone diseases such as fibrous dysplasia, ossifying and non-ossifying fibroma, osteoma and osteoid osteoma may all present with visual symptoms and may lead to visual loss due to optic nerve compression or vascular compromise. Metastatic bony lesions, particularly from prostate, breast, lung and myeloma in adults and neuroblastoma in children (Fig. 4.21), are also relatively common and may be the initial presentation of disseminated disease. A number of subperiosteal lesions are particularly common in or peculiar to the orbit and are described here.

Meningioma

Cranial meningiomas arising from the sphenoid wing or parasellar region commonly extend into the orbit (Figs 4.22 and 4.23). Meningiomas are three times more common in women and most commonly present in the fifth or sixth decades. Angioblastic and sarcomatous meningiomas are rare aggressive forms which present at an earlier age and carry a poor prognosis [5].

Plain films may demonstrate hyperostosis at the site of tumour origin which is commonly seen with orbital wall meningiomas. Bone remodelling with dilatation of adjacent air spaces (pneumosinus dilatans) or widening of the ophthalmic fissures may also be present and areas of bone destruction can be seen with more aggressive lesions. CT demonstrates a homogeneous enhancing mass lying adjacent to bone. The mass may have an *en plaque* component with extension along the periosteum and may extend through the marrow space of the affected bone to lie in both the orbit and the intracranial cavity. The presence of a thin *en plaque* extension adjacent to bone may lead to underestimation of the true extent of tumour spread if post-contrast images are not

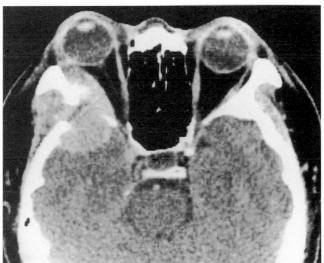

Fig. 4.22 Post-contrast CT showing bone destruction due to a meningioma of the sphenoid wing.

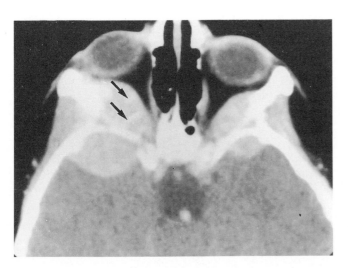

Fig. 4.21 Post-contrast CT showing metastatic neuroblastoma deposits in both sphenoid wings. Note the dystrophic bone formation within the tumour mass (arrows).

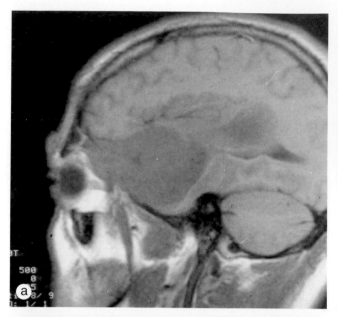

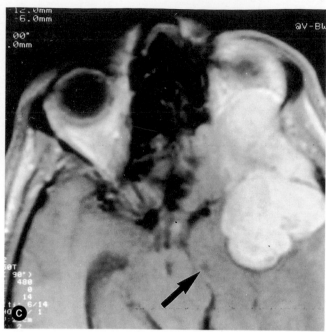

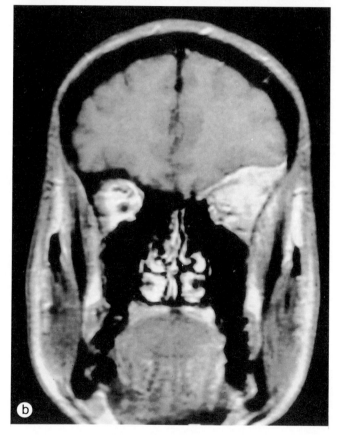

performed [2,3,5]. On MR imaging orbital mening-iomas appear isointense to muscle on T1-weighted images and hyperintense to fat on T2-weighted images (Fig. 4.23). This contrasts with the appear-ances of perioptic meningiomas (see above) which are isointense to fat on T2-weighted images. All orbital meningiomas show contrast enhancement and post-contrast images are essential to accurately determine the extent of *en plaque* and intracranial extension [1].

Dermoid cysts

Dermoid cysts are clinically divided into superficial and deep dermoids. Superficial dermoids present in infancy as rounded periorbital masses, usually situ-ated temporally where they arise from the tempero-zygomatic suture. Deep dermoids present later in life as slow-growing orbital masses. They can arise from any suture site and present with proptosis and visual disturbance. Deep dermoids often extend from the orbit into the temporal fossa, the pterygo-palatine fossa or even intracranially. Treatment is by surgical excision but this can be technically difficult or even impossible [1,2,5,15].

On CT scan superficial dermoids appear as round-ed, well-defined masses which often have a central area of fat attenuation (Fig. 4.24). Deep dermoids have well-defined margins which often contain calcification (50%), particularly when large. Most dermoids appear heterogeneous and internal fat

Fig. 4.23 (**a**) Parasagittal T1-weighted MR showing an extensive sphenoid wing meningioma extending into the left orbit. (**b**) Coronal T1-weighted MR following gadolinium showing extension of tumour through the sphenoid wing. (**c**) Transverse T1-weighted MR following gadolinium showing early tentorial herniation of the temporal lobe due to intracranial tumour (arrow).

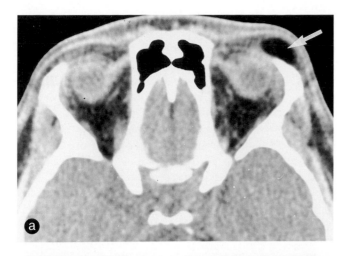

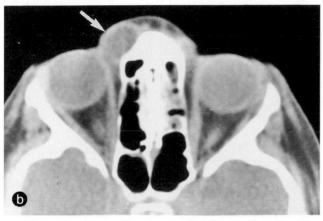

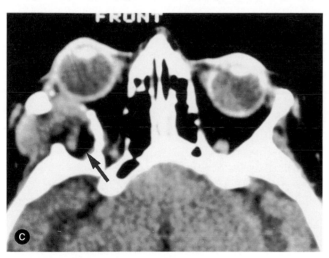

Fig. 4.24 (**a**) CT showing a fat-containing exterior angular dermoid cyst (arrow). (**b**) CT showing an internal angular epidermoid cyst (arrow). (**c**) CT showing a deep dermoid of the orbit extending through the lateral orbital wall. Note the internal fat density (arrow) and the bowing of the lateral orbital wall. ((**a**) and (**b**) reproduced by courtesy of Dr RW Whitehouse, Department of Clinical Radiology, University of Manchester.)

densities, which are seen in 30–40%, are almost pathognomonic since orbital lipomas and liposarcomas are extremely rare. Layering of different components with a fat/fluid level is also well described although unusual. Heterogeneous contrast enhancement occurs in approximately 50% although post-contrast images are often unnecessary for the diagnosis. Thinning or notching of underlying bone is seen in over 80%, reflecting the chronicity of the disease process, and clearly marginated full-thickness defects in the bone are present in 10–15% of cases. Dermoid cysts have a high signal intensity on both T1- and T2-weighted MR images, reflecting the presence of fat within them. Epidermoid cysts also occur in the orbit and are radiologically indistinguishable from dermoids although they do not contain fat and typically appear slightly hyperintense on T1-weighted images due to their protein content or minor internal haemorrhage [1].

Subperiosteal haematomas

Subperiosteal haematomas may be associated with trauma or may arise spontaneously, particularly in patients with disordered clotting. They are commonest in children and young adults and present with sudden onset of visual disturbance usually associated with proptosis. Incompletely resorbed haematomas may give rise to a chronic chocolate cyst containing altered blood and CT demonstrates a high-density, well-defined lentiform mass lesion which does not enhance with contrast. The typical signal characteristics of haematomas allows a definitive diagnosis to be made on MRI images.

Lesions of the lacrimal fossa

Most mass lesions within the lacrimal fossa originate from the gland itself. Bony metastases and external angular dermoid cysts may also arise in this location but rarely present a diagnostic problem Mucoceles of the lateral part of the frontal sinus and cholesterol granulomas of the frontal bone may also be encountered, usually occurring after previous bony injury. Lesions of the gland itself are equally divided between neoplastic and inflammatory disorders. Preoperative diagnosis is of importance particularly in the case of benign mixed (pleomorphic) adenomas where complete surgical excision has a 99% cure rate while incisional biopsy is associated with a high incidence of tumour recurrence and malignant transformation [2,5].

Infective lesions

Patients with viral and bacterial dacryoadenitis present with acute inflammation of the gland which may be suppurative. They may show systemic features of infection, preauricular and cervical adenopathy and leukocytosis on laboratory investigations. CT demonstrates diffuse gland enlargement with irregular margins and heterogeneous contrast enhancement [2,5].

Non-specific inflammation

Idiopathic dacryoadenitis is the commonest inflammatory lesion of the lacrimal gland and the third commonest site of non-specific orbital inflammatory disease [5,8]. There is a 3:1 female to male preponderance and a recognized association with systemic autoimmune disorders. The disease is bilateral in 30–40% of cases and presents with gland swelling and tenderness associated with local inflammation and pain. On CT the lesion is characteristically confined to the supero-lateral orbit with a diffuse, irregularly marginated mass which shows heterogeneous contrast enhancement with no associated calcification or bony abnormality. The mass may indent the globe and may extend into the retrobulbar fat. There may also be evidence of more multifocal disease affecting the sclera, extraocular muscles or other orbital structures. Mild cases may resolve spontaneously and more severe cases will response to steroid therapy over 1–4 months.

Specific inflammations

Dacryoadenitis may occur as a part of several systemic disorders, including Sjögren's syndrome, sarcoidosis and Wegener's granulomatosis. The clinical presentation is usually subacute or chronic and, in most cases, is associated with other evidence of the systemic diseases. CT appearances are indistinguishable from idiopathic inflammatory disease and biopsy is required to establish where there is clinical doubt.

Neoplastic lesions

Lacrimal gland tumours account for 7–8% of orbital neoplasms. Lymphomas form one-third of lacrimal gland tumours; the remainder are of epthelial origin of which half are benign pleomorphic (mixed) adenomas and the remainder are malignant [3,5].

Epithelial tumours

Benign pleomorphic adenomas are slow-growing encapsulated lesions which most commonly present between 20 and 60 years of age. They have a tendency to recur (32%), often with malignant transformation, following excisional biopsy and the lesion should be removed with the capsule intact wherever possible [3,5]. In contrast to inflammatory lesions epithelial tumours most commonly arise from the orbital lobe of the gland and therefore tend to become large before clinical presentation. CT demonstrates a well-defined oval or rounded mass lesion in the lacrimal fossa that may show rather nodular margins (Fig. 4.25). Calcification may be present within the lesion and bone erosion is common although bone destruction is not seen. These tumours exhibit moderate to marked contrast enhancement and may contain non-enhancing areas of cystic degeneration.

Carcinomas of the lacrimal gland

In many cases CT will not distinguish between benign and malignant lesions of the lacrimal gland. Evidence of bone destruction is highly suggestive of a malignant lesion, but up to 70% of malignant lesions will show only pressure erosions of bone identical to those seen with benign tumours. Malignant lesions tend to have a less rounded, oblong contour, frequently show extensive posterior extension and more often contain calcification. Irregularities in the outline of the tumour and the presence of satellite nodules are also in favour of a malignant lesion.

Lymphoma of the lacrimal gland

Orbital lymphoma occurs principally in the sixth and seventh decades of life and has a predilection for the anterior orbit. On CT lymphoma appears as a well-defined mass which follows the natural contours of orbital contents or engulfs them without displacement. In most cases lymphoma extends beyond the lacrimal gland or arises elsewhere and is discussed in more detail below.

Magnetic resonance imaging of lacrimal gland masses

MRI appearances of lacrimal gland masses are highly variable. Although epithelial neoplasms are generally of high intensity on T2-weighted images there is no apparent distinction between malignant and

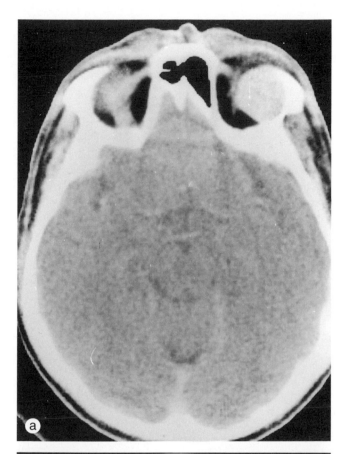

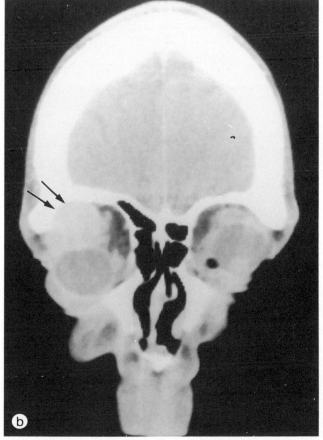

Fig. 4.25 Transverse (**a**) and coronal (**b**) post-contrast CT showing a pleomorphic adenoma of the lacrimal gland. The mass is well defined, enhances homogeneously and is associated with bone erosion (arrows).

benign lesions. Furthermore, inflammatory disorders give rise to a wide and non-specific variety of appearances, particularly on T2-weighted images [1].

Solitary orbital mass lesions

Orbital hamartomas

A hamartoma is an area of overgrowth of normal mature cells within an organ. The exact dividing line between a hamartoma and a benign tumour is ill-defined, but within the orbit most workers consider haemangiomas and lymphangiomas to be true hamartomatous lesions.

Infantile capillary haemangiomas

Infantile capillary haemangiomas result from an abnormal proliferation of blood vessels and consist of a network of vascular channels lined by normal endothelial cells. Clinically they are commoner in females, present within the first few months of life and undergo spontaneous involution by 10–15 years of age [2,3,5]. Orbital haemangiomas may lie within the preseptal or postseptal spaces and often have a large dermal component with striking facial deformity. CT demonstrates a markedly enhancing mass lesion which may have well- or poorly-defined margins. Displacement and indentation of the globe is common and there may be displacement of extraocular muscles and enlargement of the affected orbit. The MRI appearance of capillary haemangiomas is very distinctive with numerous areas of internal linear flow voids due to vascular channels, which appear as areas of high signal on T1-weighted gradient echo sequences [1,2,5]. Angiography usually demonstrates multiple feeding vessels and a prominent tumour blush with early venous opacification.

Cavernous haemangioma

Cavernous haemangiomas are the commonest benign mass lesions occurring in the orbit. They consist of large vascular channels with abundant smooth muscle and are delineated by a fine fibrous capsule. They grow slowly and usually present between 30 and 60 years of age with proptosis and minor visual disturbance. Treatment is by surgical excision and results in cure [1,2].

CT demonstrates a rounded, well-defined mass lesion which may rarely contain areas of internal calcification (Fig. 4.26). There is commonly indentation of the posterior aspect of the globe and compensatory bowing of the lateral wall of the orbit is present in 50%, reflecting the slow growth rate. Post-contrast images demonstrate a gradual increase in attenuation for up to 15 minutes after contrast injection reflecting slow internal tumour circulation

[3]. There may be internal pooling of contrast within large vascular spaces, especially on delayed images. Angiography demonstrates an apparently avascular mass but may also show small internal puddles of contrast which appear late in the arterial phase and persist late into the venous phase. Cavernous haemangiomas are isointense to extraocular muscles on T1-weighted images but may contain areas of high signal intensity due to thrombosis. They are markedly hyperintense on T2-weighted images and enhance with intravenous contrast (Fig. 4.27). They do not show evidence of internal flow void and if present this should suggest a high flow lesion such as haemangiopericytoma, arteriovenous malformation or varix.

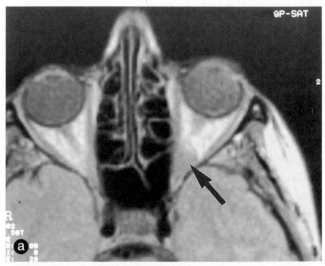

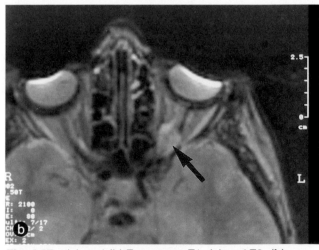

Fig. 4.27 (**a**) and (**b**) Transverse T1- (**a**) and T2- (**b**) weighted MR images showing a cavernous haemangioma in the orbital apex (arrows). Note the extensive paramagnetic artefact affecting the anterior part of both orbits (white arrows) due to mascara. (**c**) and (**d**) Parasagittal T1-weighted images before (**c**) and after (**d**) gadolinium showing compression of the optic nerve by an enhancing apical mass.

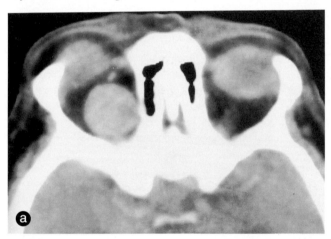

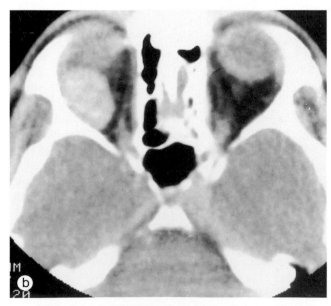

Fig. 4.26 Post-contrast CT showing intraconal and extraconal (**b**) cavernous haemangiomas.

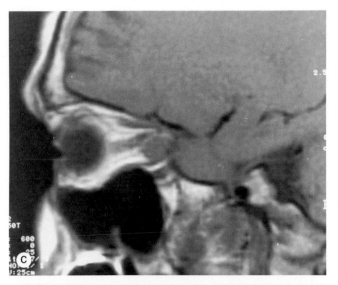

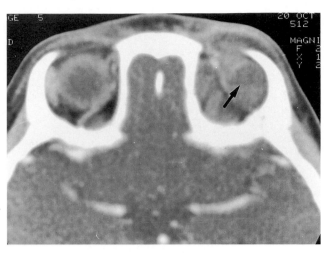

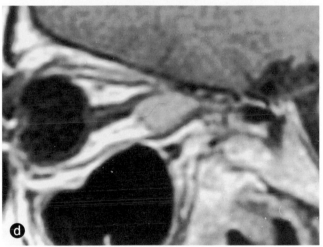

Fig. 4.28 CT showing a cystic mass in the left orbit in a patient presenting with acute onset of painful proptosis. Note the fluid/fluid level within the cyst (arrow) due to recent haemorrhage within a deep orbital lymphangioma. (Reproduced by courtesy of Dr RW Whitehouse, Department of Clinical Radiology, University of Manchester.)

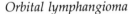

Orbital lymphangioma

Orbital lymphangiomas are complex lesions consisting of endothelially lined channels containing serous fluid, abnormal blood vessels, smooth muscle and haemorrhagic cysts. Despite their apparently vascular histology they are isolated lesions and receive only a stromal blood supply. Lymphangiomas may be superficial, retrobulbar or combined. Superficial and combined lesions present early in life and enlarge progressively over the following years. Deep lesions are often occult and usually present in early adulthood with acute onset of proptosis and optic nerve compression resulting from spontaneous internal haemorrhage. Surgical resection is complicated by the poorly defined margins and extent of the lesions [5, 16].

CT demonstrates an ill-defined soft tissue mass which may extend through several compartments. The mass often contains internal cystic areas which may be of high attenuation following recent haemorrhage or may contain a fluid/fluid level (Fig. 4.28).

Phleboliths are seen in 5–10% of cases and orbital enlargement is seen in 50%. Contrast enhancement is heterogeneous and cystic areas may show rim enhancement. Angiography demonstrates an avascular lesion or a slight blush in the late arterial phase and contrast puddling does not occur. Direct injection of contrast into the lesion has been employed to exclude a diagnosis of orbital varix. This results in filling of multiple small vascular channels and cysts which persist for longer than 10 minutes with no evidence of venous drainage. MRI typically demonstrates a heterogeneous lesion with high-signal areas on T1-weighted images due to previous internal haemorrhage and high-signal areas on T2-weighted images due to serous and haemorrhagic cysts (1,2,3,5,16] (Fig. 4.29).

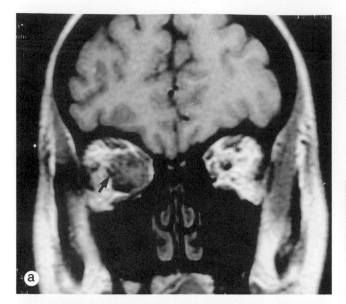

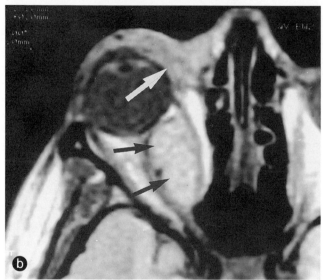

Fig. 4.29 (**a**) Coronal T1-weighted MRI in a patient with a combined deep and superficial lymphangioma. Note the displacement of the optic nerve and intact perineural space (arrow). (**b**) Transverse T1-weighted image following gadolinium showing the enhancing lymphangioma (black arrows) extending superficially to the orbital septum (white arrow).

Orbital varices

Orbital varices consist of multiple venous channels which can be difficult to distinguish histologically from lymphangioma. Unlike lymphangioma they have venous feeding and draining vessels and may distend when jugular venous pressure is raised. In most cases the varix is minimally distensible and presents in adolescence or early adulthood with recurrent episodes of acute painful proptosis due to thrombosis or haemorrhage within the lesion [2,5].

CT demonstrates a markedly enhancing, well-defined soft tissue mass which may be superficial or deep to the globe. Internal calcification due to phlebolith formation is particularly common in distensible varices and internal haemorrhage may also be seen. Repetition of the examination during a Valsalva manoeuvre or with the patient in the prone position may demonstrate an increase in the size of the mass even in patients with clinically non-distensible varices (Fig. 4.30). Where this sign is absent

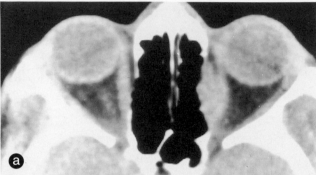

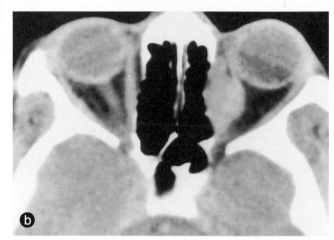

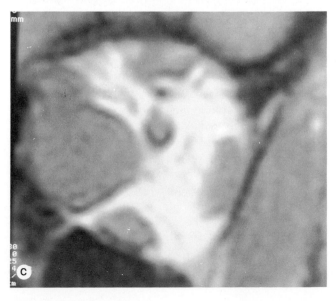

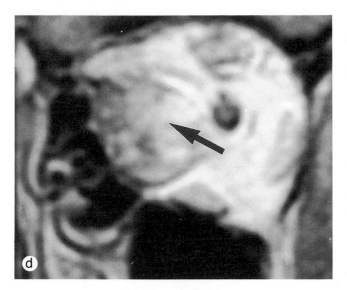

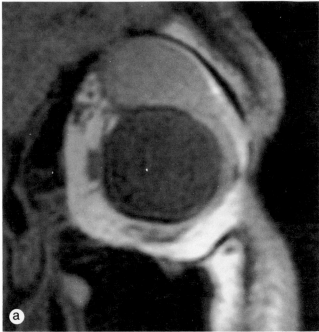

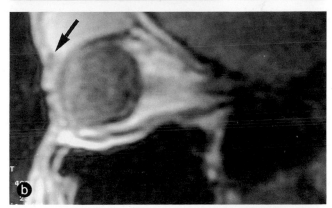

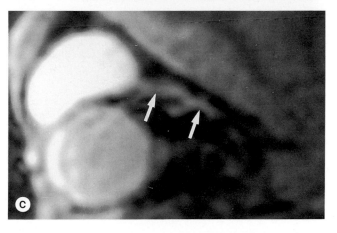

Fig. 4.30 (**a**) and (**b**) CT images demonstrating orbital varices arising from the medial orbital wall. Note the increase in size of the mass during jugular vein compression (**b**). (**c**) and (**d**) Coronal T1-weighted MR images before (**c**) and after (**d**) gadolinium. On the precontrast image the mass is isointense to extraocular muscles. Following gadolinium the mass shows heterogeneous enhancement (arrow).

it is not possible to distinguish between orbital varix and lymphangioma on CT findings alone. Direct injection of contrast into the lesion or orbital venography will demonstrate a prominent blush with venous drainage. The MRI appearances are of a mass lesion which may demonstrate flow-related signal void but which is more usually of intermittent signal intensity on T1-weighted images. It is important to realize that even lesions with distended vessels containing stagnant blood seldom show evidence of paramagnetic contents since oxygenated blood is diamagnetic. Areas of thrombosis will appear as high-signal areas on T1-weighted images and may extend into draining vessels. T2-weighted sequences demonstrate a high signal within the mass and the use of T2 or STIR images may allow the identification of distended draining veins [1,2,3,5] (Fig. 4.31).

Fig. 4.31 (**a**) Coronal T1-weighted MRI showing traumatic varices following a stab wound to the eye. The mass is displacing the globe downwards. (**b**) Parasagittal T1-weighted MR image showing anterior extension of the mass into the superior fornix at the site of injury (arrow). (**c**) Parasagittal STIR image clearly demonstrates the distended draining vein (arrows).

Tumours arising from vascular tissues

Haemangiopericytoma

These are rare vascular tumours arising from the pericytes of Zimmerman which normally surround capillaries. They are commonest in the fourth decade and half are metastatic in origin although metastatic spread from the orbit is rare. Recurrence following excision is very common (30% by 5 years). CT demonstrates an enhancing mass most commonly within the intraconal space; however, unlike cavernous haemangiomas, the margins of the lesion are usually poorly defined. Bone destruction may be demonstrated but calcification does not occur. Angiography demonstrates a prominent early blush without the venous contrast pooling seen in cavernous haemangiomas.

Tumours arising from peripheral nerves

Peripheral nerve tumours comprise 4% of orbital neoplasms. Half of these are plexiform neurofibromas with the remaining half divided equally between solitary neurofibromas and schwannomas [5,17].

Plexiform neurofibroma

Plexiform neurofibromas are pathognomonic of type 1 neurofibromatosis. They are unencapsulated, highly vascular and grow both along peripheral nerves and into surrounding tissues. Clinically they present early in life, usually before 10 years of age, with thickening of the periorbital soft tissues and lid. The dermal extension of the tumour is often extensive and ocular abnormalities such as corneal Lisch nodules and choroidal hamartomas are present in most. Surgical management is unsatisfactory and the results are usually temporary [5,18].

Plain film changes are often striking with compensatory enlargement of the orbit combined with primary mesodermal defects. Common abnormalities consist of widening of the superior and inferior orbital fissures, hypoplasia of the ethmoid and maxillary sinuses, defects of the greater wing of the sphenoid and of the orbital roof, elevation of the lesser wing of the sphenoid and enlargement of the middle cranial fossa.

CT will demonstrate these bony changes together with the extent of the tumour which appears as an ill-defined enhancing soft tissue mass. Thickening of the lids and anterior periorbital tissues with loss of the fat planes is seen in most patients. Enlargement of the extraocular muscles is common and involve-

ment of the ciliary nerves may lead to irregular thickening of the optic nerve sheath. Orbital fat is often of rather increased attenuation which is believed to represent involvement of small nerves. Early onset of glaucoma causes compensatory expansion of the globe (buphthalmos) in up to 50% of cases, but unlike other causes of buphthalmos posterior scleral thickening and enhancement are also seen [18]. On MR images the diagnosis depends on the same combination of morphological features. The tumours appear very similar to lymphangiomas, but the clinical setting and associated abnormalities usually allow a confident diagnosis.

Solitary neurofibroma and schwannoma

Solitary neurofibromas are benign lesions most commonly arising from branches of the fifth nerve. They present in middle age with proptosis and symptoms related to mass effect. Schwannomas may arise from any peripheral nerve, present between 20 and 50 years of age and are associated with neurofibromatosis in 18% of cases. In comparison to neurofibromas they tend to be larger and to occur more commonly in the inferior quadrants and in the intraconal space. Solitary neurofibroma and schwannoma are indistinguishable on all imaging modalities [1,17] and appear as well-defined enhancing masses on CT. On MRI they are isointense to muscle on T1-weighted images and hyperintense on T2 and post-contrast images (Fig. 4.32).

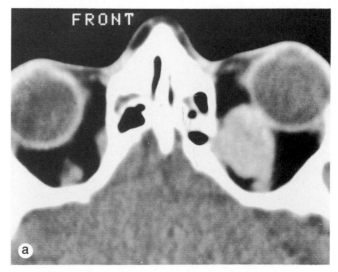

Fig. 4.32 (**a**) CT showing a schwannoma presenting as an enhancing mass in the left orbit. (**b**) T1-weighted MR image showing a well-defined mass, isointense to muscle, extending into the extra- and intraconal spaces. (**c**) T1-weighted image showing tumoral enhancement following gadolinium (arrow).

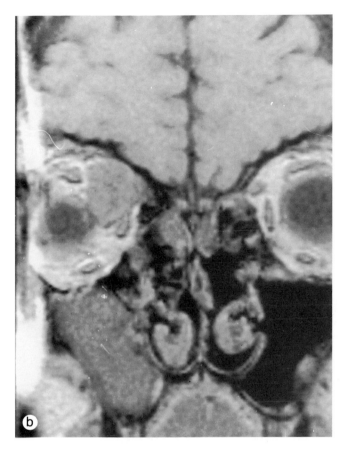

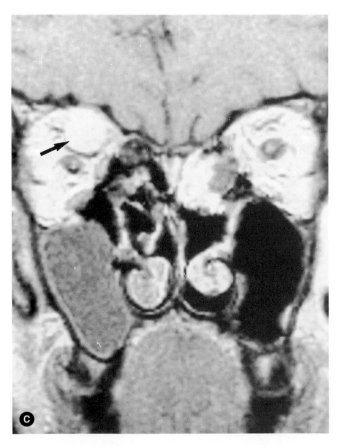

Malignant degeneration may rarely occur in schwannomas and is usually associated with neurofibromatosis. Loss of tumour delineation on CT or MR images, together with rapid growth or bone destruction, should arouse suspicion of malignant degeneration.

Tumours arising from muscle

Rhabdomyosarcoma

Rhabdomyosarcoma is the commonest primary malignant tumour of the orbit in children. Although it has been reported from birth to the age of 70, over 75% present before the age of 10. The tumour is commoner in males (3:2) and presents with rapidly progressive proptosis, inflammatory changes in the periorbital tissues and pain. Orbital involvement may also occur due to spread of tumours arising in the sinuses or skull base.

The traditional histological classification into embryonic, alveolar and pleomorphic tumours relates poorly to prognosis and a new system has been devised which divides rhabdomyosarcoma into three types more closely related to clinical behaviour. Anaplastic and monomorphous types jointly constitute 20% of cases and carry a poor prognosis while rhabdomyosarcoma (MX) consists of a mixture of subtypes all associated with a better outcome. Orbital tumours tend to be of low grade and have a particularly favourable prognosis, with a 90% 5-year survival rate [5,19]. Initial investigation is aimed at documenting the extent of metastatic spread which, with orbital tumours, is most commonly to the lungs and bone.

CT demonstrates a large soft tissue mass, most commonly in the upper nasal quadrant. The mass may demonstrate internal heterogeneity with areas of low attenuation. Post-contrast enhancement is normal and may also be strikingly heterogeneous. Calcification does not occur but internal haemorrhage may be seen in aggressive tumours. On MRI the tumour is isointense to muscle on T1-weighted images and hyperintense to fat with T2 weighting [1]. Orbital extension of extraorbital rhabdomyosarcoma is invariably associated with extensive bone destruction and with mucosal and meningeal spread. Meningeal spread is particularly well seen as areas of meningeal thickening and enhancement on post-contrast MR images. This form of tumour spread carries a poor prognosis but is very rare with isolated orbital tumours.

Tumours arising from lymphatic tissue

Lymphoid tumours account for 10–15% of orbital masses and include a wide spectrum of diseases from malignant lymphoma to reactive and atypical hyperplasia. All these disorders are characterized by lymphocyte proliferation, a tendency to recur after treatment and an association with disease in extraorbital sites. Although the histological classification of lymphoproliferative disorders is complex their clinical behaviour and radiological appearances can be appreciated by consideration of three groups of disease [5].

Reactive lymphoid hyperplasia: A benign disorder presenting with a slowly growing orbital mass. These masses are almost always anterior and often extend into the subconjunctival space. Most patients present in their fourth or fifth decade. Some patients have local pain and the majority will respond to steroids although a minority require single-agent chemotherapy or low dose radiotherapy.

Atypical lymphoid hyperplasia: These lesions range from apparently benign to more aggressive lesions which are clinically indistinguishable from low-grade lymphomas. Approximately 30% of patients have bilateral disease, 30% have disseminated extraocular disease and up to 25% will develop malignant transformation. Most patients are resistant to steroid therapy and require combination chemo- and radiotherapy.

Lymphoma: Lymphoma forms the largest group of lymphoproliferative disorders affecting the orbit. All primary orbital tumours are non-Hodgkin's lymphomas and high-grade tumours are very rare. Between 40% and 70% will have evidence of systemic lymphoma. The onset of symptoms is classically in the sixth and seventh decades with a palpable mass in the anterior orbit which is often visible below the conjunctiva. Treatment depends on the subtype of lymphoma and stage of disease and initial investigations include biopsy and a full staging workup.

Radiological features

In most cases CT demonstrates a very well-defined, homogeneous, mildly enhancing mass in the anterior orbit. The mass is often rather dense on precontrast scans although calcification is not seen prior to therapy. There is usually an intraorbital, extraconal component which may spread intraconally, but isolated intraconal lesions are unusual (10%). Most lymphoid masses arise in the upper outer quadrant

and many involve the lacrimal gland (see above). The lesions are often bilateral (25%) and displacement of the globe is common. Lymphoproliferative tumours typically mould to the contours of the globe and orbital contents and this feature alone is highly suggestive of the diagnosis (Fig. 4.33). Although

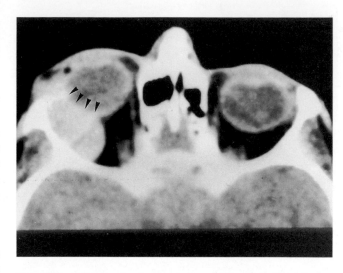

Fig. 4.33 CT showing a lymphoma of the orbit. Note that the mass conforms to the outlines of the globe (arrowheads).

bone destruction may be seen in patients with malignant lymphoma bone remodelling is rare and should cast doubt on the diagnosis [1,2,20,21]. Reactive lymphoid hyperplasia is not reliably distinguished from lymphoma on imaging criteria although the diagnosis is suggested if the mass appears heterogeneous on pre- and post-contrast images [22]. On MR images lymphoproliferative lesions may be indistinguishable from idiopathic orbital inflammatory syndrome (see above) since they are commonly hypointense to orbital fat on T2-weighted images.

Orbital metastases

Metastatic deposits in the orbit most commonly present in patients with known primary disease but can be the first manifestation of disease. CT demonstrates ill-defined masses which usually enhance with contrast and which most commonly lie within the intraconal space. Multifocal disease is common and rapid growth may lead to a 'frozen orbit' with loss of delineation of orbital soft tissues (Fig. 4.34).

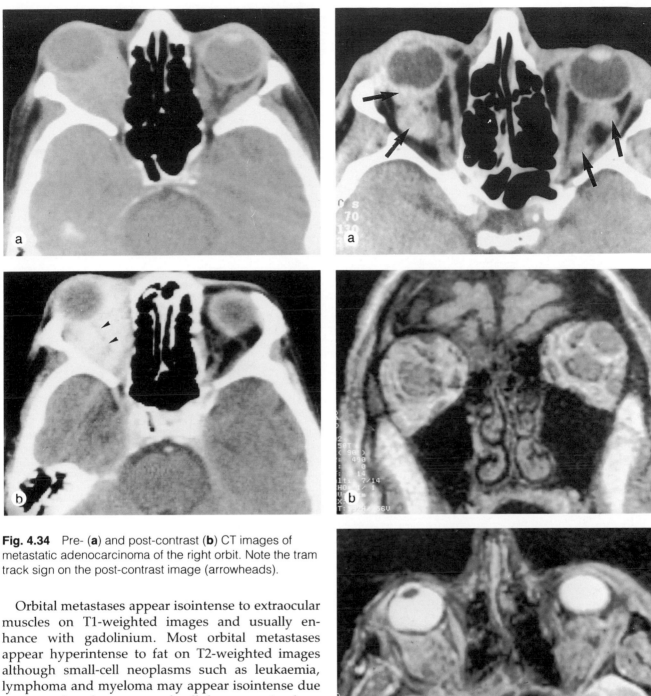

Fig. 4.34 Pre- (**a**) and post-contrast (**b**) CT images of metastatic adenocarcinoma of the right orbit. Note the tram track sign on the post-contrast image (arrowheads).

Orbital metastases appear isointense to extraocular muscles on T1-weighted images and usually enhance with gadolinium. Most orbital metastases appear hyperintense to fat on T2-weighted images although small-cell neoplasms such as leukaemia, lymphoma and myeloma may appear isointense due to the reduced amounts of cellular water in these diseases (Fig. 4.35).

Inflammatory disorders of the orbit

Inflammatory disorders of the orbit typically involve several orbital compartments. They should be suspected in the presence of diffuse or particularly ill-defined imaging abnormalities and where the clinical presentation is acute in onset, painful or associated with conjunctival injection or systemic fever and malaise.

Fig. 4.35 (**a**) Post-contrast CT showing multiple enhancing soft tissue orbital masses due to multiple myeloma (arrows). (**b**) T1-weighted coronal image showing myeloma deposits isointense to extraocular muscle. (**c**) T2-weighted image. Myeloma deposits are isointense to orbital fat.

Orbital pseudotumour

The term 'orbital pseudotumour' has been commonly applied to inflammatory mass lesions occurring within the orbit. Despite its common use the term is ill defined and has been used by some workers to describe all non-neoplastic mass lesions, by others to describe masses responsive to steroid therapy and by others as a synonym for the group of idiopathic orbital inflammatory disorders described below. The term is still in common clinical usage largely due to the practice of treating suspected idiopathic orbital inflammatory syndrome with steroids in the absence of biopsy proof. Since many inflammatory and lymphoid disorders will respond to blind therapy the term 'pseudotumour' is often applied to this very heterogeneous group. In general the term is falling into disfavour since its indiscriminate use may lead to inappropriate management in individual cases.

Acute and subacute inflammatory disorders

Orbital cellulitis

Microbial cellulitis is a major cause of orbital inflammation occurring most commonly in children and young adults. Treatment is a matter of urgency since the condition is sight-threatening and can result in cavernous sinus thrombosis and cerebral abscess formation. In adults and older children cellulitis most commonly occurs secondary to sinus disease or as a complication of trauma or dental extraction and surgical drainage of the sinuses or of a subperiosteal abscess is required in most cases. Complications are relatively common in this group and include permanent loss of vision, neurological impairment and osteomyelitis.

Plain films may demonstrate the extent of sinus disease but are incapable of providing adequate information in most cases. Patients with localized preseptal inflammation do not require CT unless there is clinical suspicion of extension to the postseptal space. The development of postseptal cellulitis is associated with a minimal increase in the attenuation of orbital fat on the affected side (5–10 HU) which is poorly demonstrated on CT. Proptosis due to swelling of inflamed orbital fat may also be present and, although this is invariably slight, it should be considered as definite evidence of postseptal involvement [23]. Subperiosteal infection produces a well-defined lentiform mass adjacent to the bony orbital wall (Fig. 4.36). Over 50% of these masses occur secondary to ethmoiditis and are

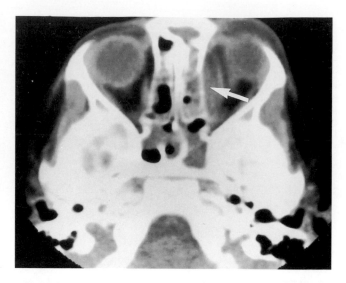

Fig. 4.36 CT showing a subperiosteal abscess along the medial wall of the left orbit (arrow). Note the soft tissue opacification of the adjacent ethmoid air cells.

located along the medial wall. In patients with anterior ethmoidal or frontal sinus disease the mass may arise along the orbital roof and may be overlooked in up to one-third of patients if coronal images are not obtained [24,25] (Fig. 4.37(c)). Extraocular muscles adjacent to the mass are commonly displaced and may be enlarged and ill defined even in the absence of significant intraorbital infection.

Rupture of a subperiosteal abscess or direct spread may result in orbital abscess formation. Orbital abscesses are most commonly extraconal and an isolated intraconal abscess should raise the suspicion of an orbital foreign body. Abscesses appear as ill-defined enhancing mass lesions. Enhancement may be homogeneous, heterogeneous or ring-like. Involvement of the globe with scleral thickening and enhancement is common.

Venous thrombosis secondary to orbital infection is common and cavernous sinus thrombosis should be suspected in any patient where orbital infection is associated with neurological deficit or rapidly increasing proptosis. CT will show enlargement of the superior ophthalmic vein and cavernous sinus with failure of enhancement following intravenous contrast administration (Fig. 4.37).

Magnetic resonance imaging can clearly demonstrate orbital infection providing similar morphological information to CT. Inflammatory soft tissue masses appear as high signal areas on T2-weighted images due to local oedema. MRI appears considerably superior to CT in demonstrating oedema in orbital fat which appears as a decrease in signal on T1-weighted sequences.

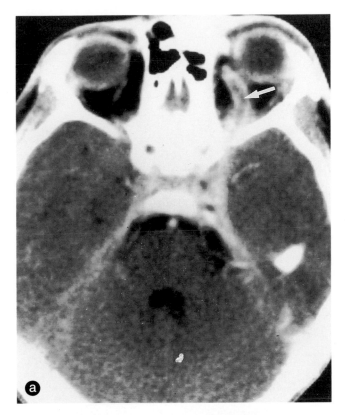

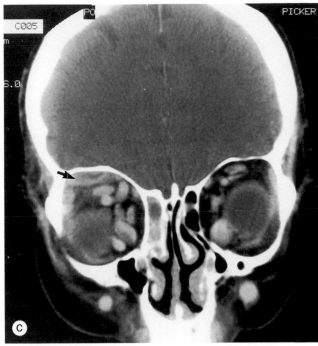

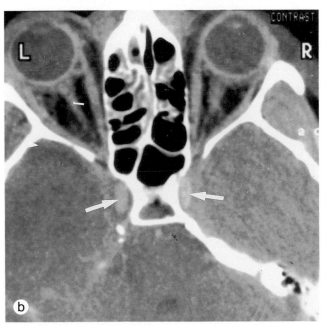

Fig. 4.37 (**a**) Post-contrast CT showing a filling defect in the superior ophthalmic vein (arrow) due to thrombosis secondary to orbital cellulitis. (**b**) Post-contrast CT showing bilateral cavernous sinus thrombosis. Note the enhancement of the intracavernous carotid arteries (arrows). (**c**) Coronal post-contrast CT showing a subperiosteal abscess of the orbital roof (arrow). Note the multiple dilated orbital veins due to cavernous sinus thrombosis. (Reproduced by courtesy of Dr T Jaspan.)

Idiopathic orbital inflammatory syndrome

Idiopathic orbital inflammatory syndrome (IOIS) describes a range of clinical syndromes characterized by acute orbital inflammation of unknown aetiology. In the adult population IOIS is the commonest cause of an orbital mass lesion and has traditionally been included within the group of lesions known as orbital pseudotumours. Treatment is with systemic steroids and the response is usually dramatic although relapse is common following steroid withdrawal. IOIS is divided into five arbitrary clinical categories on the basis of the anatomical site of involvement. The myositic and lacrimal forms of IOIS have been described above, the three remaining categories [2,5,26] are described below.

Idiopathic anterior orbital inflammation: Inflammation limited to anterior structures presents with pain, proptosis, lid swelling and decreased visual acuity. Scleritis is frequent and may cause exudative retinal detachment. CT demonstrates diffuse anterior orbital infiltration with lid swelling and displacement of the globe (Fig. 4.38). The globe is invariably involved, with scleral and choroidal thickening obscuring the insertion of the optic nerve. All the inflammatory tissue is poorly delineated with involvement of adjacent fat and shows moderate to marked enhancement.

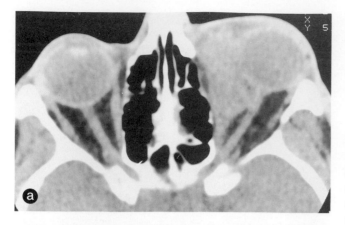

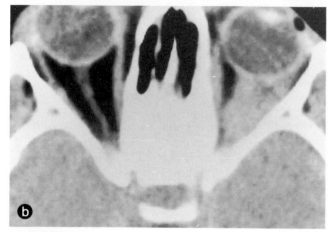

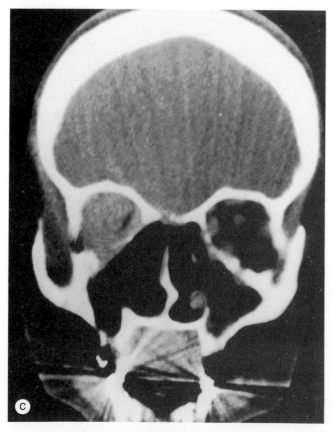

Fig. 4.38 (**a**) Post-contrast CT showing an ill-defined enhancing mass in the left orbit. Diagnosis: idopathic anterior orbital inflammation. Transverse (**b**) and coronal (**c**) CT images show an idiopathic apical inflammatory mass engulfing the rectus muscles and optic nerve.

Idiopathic diffuse orbital inflammation: This unusual form of IOIS results from inflammation involving the entire orbit occurring principally in adolescence and early childhood. CT demonstrates abnormal soft tissue extending throughout the orbit as far as the apex. The response to steroids is good but relapse is common on treatment withdrawal.

Idiopathic apical inflammation: Inflammation at the orbital apex gives rise to prominent loss of function in the presence of minimal inflammatory signs. Patients present with pain, decreased visual acuity and minimal proptosis. CT demonstrates an irregular enhancing soft tissue mass infiltrating the orbital apex. Extension along the optic nerve and involved extraocular muscles are common.

Magnetic resonance imaging: MRI demonstrates similar morphological changes to CT in IOIS but provides some increase in the specificity of diagnosis. Inflammatory masses appear isointense to normal extraocular muscle on T1-weighted sequences and minimally hypo- or hyperintense on T2-weighted images. This allows distinction from most other malignant processes which have a high signal, equal to or greater than orbital fat on T2-weighted images. A number of neoplastic lesions (myeloma, lymphoma and neuroblastoma metastases) may have similar appearances to IOIS, with a relatively low signal intensity on T2-weighted images which is believed to reflect the hypercellular nature of these tumours.

Chronic inflammatory disorders

Idiopathic sclerosing orbital inflammation

This rare form of pseudotumour is characterized by orbital fibrosis and may be associated with systemic disorders including retroperitoneal fibrosis, Peyronie's disease and Dupytren's contracture. The response to steroids is poor and orbital radiotherapy should be used as soon as the diagnosis is confirmed. CT is unable to distinguish this entity from IOIS and MRI appearances have not been described. Persistence of CT abnormality following steroid therapy in patients with IOIS, together with the clinical features should lead to suspicion.

Tolosa-Hunt syndrome

The Tolosa-Hunt syndrome consists of painful ophthalmoplegia caused by cavernous sinus inflammation. Clinically there is gnawing retro-orbital pain lasting days to weeks with occasional spontaneous remissions. Deficits in the third, fourth, sixth or first branch of the fifth cranial nerves are common and involvement of the optic nerve or the sympathetic fibres around the internal carotid artery have been described. The response to steroid therapy is rapid and dramatic. CT may demonstrate asymmetric enlargement of the cavernous sinus due to an enhancing soft tissue infiltrate which may extend through the superior orbital fissure into the orbital apex. On MRI the mass is isointense to fat on T2-weighted images and isointense to extraocular muscles on T1-weighted images with prominent enhancement following intravenous contrast. Orbital venography will demonstrate occlusion of the superior ophthalmic vein at the orbital apex without displacement which may help in the differentiation between Tolosa-Hunt syndrome and apical malignancy (Fig. 4.39).

Other chronic inflammatory disorders

Ocular or orbital involvement is common in several systemic inflammatory disorders. Sarcoidosis is associated with uveitis, keratoconjunctivitis or chorioretinitis and with lacrimal gland enlargement in 7% of patients. Involvement of orbital soft tissues is very rare and is invariably associated with CNS disease. Ocular or orbital manifestations occur in 40–50% of patient with Wegener's granulomatosis, with ill-defined enhancing mass lesions and lacrimal gland enlargement. Concomitant sinus disease is present in almost all and bone destruction may be demonstrated on plain films or CT.

ACKNOWLEDGMENTS

The authors would like to thank Mr J Yates for producing the illustrations for this chapter and Mrs J Ford for typing the manuscript.

REFERENCES

1. Atlas SW, Galetta SI. The orbit and visual system. In: Atlas SW, ed. Magnetic resonance imaging of the brain and spine. New York: Raven Press, 1991.
2. Mafee M, ed. Imaging in ophthalmology, part 1. Radiol Clin N Amer 1987; 25 (3).
3. Mafee M, ed. Imaging in ophthalmology, part 2. Radiol Clin N Amer 1987; 25 (4).
4. Berges O, Vignaud J, Aubin ML. Comparison of sonography and computed tomography in the study of orbital space-occupying lesions. Am J Neuroradiol 1984; 5: 247–251.
5. Rootman J. Diseases of the orbit: a multidisciplinary approach. Philadelphia: Lippincott, 1988.
6. Sarkies NJC. Optic nerve sheath meningioma: diagnostic features and therapeutic alternatives. Eye 1987; 1: 597–602.
7. Johns TT, Citrin CM, Black J, Sherman JL. CT evaluation of perineural orbital lesions: evaluation of the 'tram-track' sign. AJNR 1984; 5: 587–590.
8. Healy R. Computed tomographic evaluation of metastases to the orbit. Ann Ophthalmol 1983; 15: 1026–1102.
9. Flanders AE, Mafee MF, Rao VM, Choi KH. CT characteristics of orbital pseudotumours and other orbital inflammatory processes. J Comp Assist Tomog 1989; 13: 40–47.

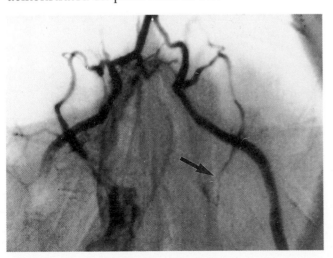

Fig. 4.39 Orbital venogram showing occlusion of the left superior ophthalmic vein at the orbital apex in a patient with Tolosa–Hunt syndrome. Note that the occluded vein is not displaced.

10. McNicholas MM, Power WJ, Griffin JF. Idiopathic inflammatory pseudotumour of the orbit: CT features correlated with clinical outcome. Clin Radiol 1991; 44: 3–7.

11. Barrett L, Glatt HJ, Burde RM, Gado MH. Optic nerve dysfunction in thyroid eye disease: CT. Radiology 1988; 167: 503–507.

12. Dresner SC, Rothfus WE, Slamovits TL, Kennerdell JS, Curtin HD. Computed tomography of orbital myositis. Am J Roentgenol 1984; 143: 671–674.

13. Nugent RA, Belkin RI, Neigel JM, Rootman J, Robertson WD, Spinelli J, Graeb DA. Graves orbitopathy: correlation of CT and clinical findings. Radiology 1990; 177: 675–682.

14. Rothfus WE, Curtin HD. Extraocular muscle enlargement: a CT review. Radiology 1984; 151: 677–681.

15. Nugent RA, Lapointe JS, Rootman J, Robertson WD, Graeb DA. Orbital dermoids: features on CT. Radiology 1987; 165: 475–478.

16. Graeb DA, Rootman J, Robertson WD, Lapointe JS, Nugent RA, Hay EJ. Orbital lymphangiomas: clinical radiological and pathological characteristics. Radiology 1990; 175: 417–421.

17. Dervin JE, Beaconsfield M, Wright JE, Moseley IF. CT findings in orbital tumours of nerve sheath origin. Clin Radiol 1989; 40: 475–479.

18. Reed D, Robertson WD, Rootman J, Douglas G. Plexiform neurofibromatosis of the orbit: CT evaluation. AJNR 1986; 7: 1259–1263.

19. Maurer JK, Foulkes M, Gehan E. Intergroup rhabdomyosarcoma study (IRS) II: preliminary report. Proc Amer Soc Clin Oncol 1983; 2: 70.

20. Mills P, Parsons CA. Primary orbital lymphoma: staging by computed tomographic scanning. Br J Radiol 1989; 62: 287–289.

21. Hornblass A, Jakobiec FA, Reifler DM, Mines J. Orbital lymphoid tumours located predominantly within extraocular muscles. Ophthalmology 1987; 94: 688–697.

22. Westacott S, Garner A, Moseley IF, Wright JE. Orbital lymphoma versus reactive lymphoid hyperplasia: an analysis of the use of computed tomography in differential diagnosis. Br J Ophthalmol 1991; 75: 722–725.

23. Towbin R, Han BK, Kaufman RA, Burke M. Postseptal cellulitis: CT in diagnosis and management. Radiology 1986; 158: 735–737.

24. Handler LC, Davey IC, Hill JC, Lauryssen C. The acute orbit: differentiation of orbital cellulitis from subperiosteal abscess by computerized tomography. Neuroradiology 1991; 33: 15–18.

25. Langham-Brown JJ, Rhys-Williams S. Computed tomography of acute orbital infection: the importance of coronal sections. Clin Radiol 1989; 40: 471–474.

26. Rowe LD, Tsiaras WT, Nichols C, Myers AR. Computerized axial tomography in inflammatory pseudotumour of the orbit. Otolaryngol Head Neck Surg 1980; 88: 378–383.

5 The paranasal sinuses

W. St C. Forbes

IMAGING METHODS

Prior to the introduction of computed tomographic (CT) scanning, pathological processes extending beyond the bony confines of the paranasal sinuses were difficult to demonstrate on imaging. Plain film techniques were supplemented with conventional tomographic techniques using complex motion tomography. Tomography carried out in different planes improved the visualization of the bony destruction and had a role in demonstrating the presence of a soft-tissue mass in the nasal cavity in cases of opaque sinus, particularly the maxillary antrum. Associated destruction of the skull base and orbit was also well demonstrated. Complex motion tomography has now been completely superseded by digital cross-sectional imaging. The current imaging methods used to evaluate sinus pathology are as follows.

Plain film techniques

Examination of the sinuses should always be made in the erect position with a horizontal X-ray beam to allow demonstration of fluid levels within the sinus. The routine conventional radiographic projections are as follows.

Occipito-mental projection (Water's view)

This view shows the maxillary antra free of overlap of the petrous bones; the sphenoid sinuses and nasopharynx can be seen through the open mouth. It is used as a survey view of the paranasal sinuses for infection, allergy, tumour and trauma.

Occipito-frontal projection (Caldwell view)

This projection demonstrates the fine detail of the frontal sinuses, orbits and nasal cavity. The lateral walls of the antra are also seen, although the overlapping petrous bones largely obscure the antra. Ethmoid sinuses are also well seen. This view is used as a survey view for the frontal and ethmoid sinuses, nasal cavity and for the detection of erosion of the lateral wall of the maxillary antrum.

Lateral projection

In this view the frontal, maxillary and sphenoid sinus are shown superimposed and it is best for showing the nasal cavities and nasopharynx.

It is used for examination of the posterior choanae for polyps, examination of the sphenoid sinuses, assessment of the bone thickness in the walls of the frontal sinus, and detection of cysts in the floor of the maxillary antra.

Submentovertical (axial) projection

This projection demonstrates the sphenoid sinuses, and also the ethmoid air cells, maxillary antra and the orbital walls. The nasopharynx is also well demonstrated. This view is used in the assessment of posterior spread of sinus malignancy.

Plain films tend to underestimate the extent of sinus disease, particularly of the ethmoids, and some centres will use CT rather than radiography as the initial screening examination.

Computed tomography (CT)

High resolution CT provides soft-tissue detail and bone detail. A thin section axial scan, with the infraorbital-meatal line parallel to the slice plane, comprises the standard plane of scanning and is the most comfortable for the patient to maintain, although in inflammatory disease the coronal plane is preferred initially.

Thin section coronal scanning allows visualization of structures lying parallel to the transverse plane that are not seen on axial scans (i.e. the palate, alveolar ridges, skull base, floors of the orbits) as well as supplementing the information obtained in the axial sections. This is particularly the case with regard to structures lying in the supero-inferior axis, e.g. medial orbital walls, nasal cavities, bony margins of the frontal, maxillary and sphenoid sinuses. To obtain this view the patient is placed in the prone position with the neck extended. Coronal sections may also be obtained in the supine, 'hanging head' position. Elderly patients and patients with rigid necks cannot be scanned in this plane. To avoid artefacts produced by dental fillings the gantry should be tilted so that the slices avoid the teeth. However this results in slices that are not in a true coronal plane.

Use of intravenous contrast

Intravenous contrast enhancement is necessary to assess soft-tissue lesions in the paranasal sinuses.

1. High dose dynamic scanning is necessary to evaluate soft-tissue abnormalities. Vascular lesions enhance with contrast and this method is best for demonstrating intracranial extension of lesions.

2. Contrast enhancement will also assist:
   (a) differentiation of a cystic, infective or necrotic portion of a lesion from a solid mass;
   (b) differentiation of fluid from solid tissue in a sinus;
   (c) evaluation of orbital involvement;
   (d) characterization of mucosal walls [1].

The main limitation of thin section CT is the partial volume effect. The thin bony walls of the antra may appear eroded, e.g. the lateral wall of the antrum in a normal patient.

Magnetic resonance imaging (MRI)

MRI provides a means of obtaining direct coronal, axial and sagittal scans. Unlike CT there is no significant image degradation from dental fillings. In view of its increased soft-tissue discrimination MRI demonstrates soft-tissue structures more clearly without the routine need for a contrast agent. Most disease processes can be differentiated, e.g. infection versus tumour versus haemorrhage.

For optimum tissue characterization T1- and T2-weighted sequences are required. Spin-echo sequences using a long echo time will give maximum T2 weighting and signal differentiation between tissues, depending upon the relaxation time of the tissues concerned. T1 weighting can be achieved using a spin-echo sequence with a short repetition time or a gradient-echo sequence, producing images of good anatomical detail.

Tissue characterization

New growths of the paranasal sinuses, whether epithelial or mesenchymal, produce signal of intermediate intensity on T1-weighted sequences and intermediate to high signal on T2 images. In contrast, retained secretions produce low signal on T1 and high signal on T2 sequences. Differentiation between tumour and retained secretion is particularly striking on T2-weighted images, the retained secretion always giving a higher signal than tumour. Furthermore, the signal given by tumour is heterogeneous due to the tumour vascularity whereas that shown by retained secretion is invariably homogeneous.

Use of paramagnetic contrast agents

Paramagnetic contrast agents produce both T1 and T2 shortening of those tissues in which they accumulate. The result on T1-weighted images is to produce an increased signal intensity of the tissue in which the gadolinium accumulates. This concentration of gadolinium is related strictly to the blood supply to that tissue, and some sinonasal tumours and many acute inflammatory processes have increased blood supply. The gadolinium enhancement is useful in obtaining more accurate assessments of tumour margins and its use is more accurate than T2-weighted images in determining true tumour margins. However, the enhancement of tumours is variable with some tumours showing no enhancement whereas most acute inflammatory processes show enhancement. Retained secretion, however, does not enhance after gadolinium. This assists in the differentiation between tumour and retained secretions producing an opaque sinus.

In summary, tumour tissue can be recognized by its heterogeneous signal; thickened mucosa, polyps and retained secretion always present homogeneous signal with or without contrast enhancement. Tumours normally have some degree of signal heterogeneity and this is increased after paramagnetic contrast, especially on T1-weighted sequences.

The main disadvantage of MRI is its inability to demonstrate bone detail which appears as signal void and may therefore be confused with other causes of signal void, e.g. dystrophic calcification, scar tissue and flowing blood. Bone is more accurately assessed by CT.

ANATOMY

The paranasal sinuses are named after the bones in which they lie and consist of the frontal, ethmoid, maxillary and sphenoid sinuses. They develop as invaginations of the mucous membrane of the nasal cavities into the adjacent bones. The mucosa is therefore similar to that found in the nasal cavity – ciliated columnar epithelium with mucus-secreting glands. The cilia are more numerous nearer the ostia of the sinuses and promote drainage of the secretions.

The paranasal sinuses are not visible at birth.

Frontal sinus

The frontal sinus develops within the frontal bone as paired cavities. They arise from the region of the frontal recess of the nose and are, in effect, displaced anterior ethmoid cells. The frontal sinuses are rudimentary at birth and are not visible radiographically until the age of 2 years, at which time they have begun to extend into the vertical plate of the frontal bone. The final adult proportions are reached only after puberty. Their development is quite variable; both fail to develop in 4% of the population.

They vary in size and shape and the cavities are usually asymmetrical in the same individual. They are frequently undeveloped on one side. A septum usually divides the two cavities into right and left compartments. Multiple incomplete septations can be present, especially in the superior portions of the sinus.

The frontal sinus has a variable drainage, reflecting its variable site of origin. Each frontal sinus drains independently through its nasofrontal duct. This duct passes through the anterior ethmoid cells and drains into the middle meatus. Less often the frontal sinus will drain directly into the anterior ethmoid air cells which in turn open into the infundiblum or on the bulla ethmoidalis.

Ethmoid sinus

The ethmoid sinuses develop and reach maturity at puberty. They are situated in the lateral masses of the ethmoid bone between the orbits and the nasal cavity. The ethmoid air cells 'labyrinth' vary in number from three to 18 on each side and are paired. Each is divided into anterior, middle and posterior cell groups according to the location of their ostia. Less commonly cells may extend into the maxillary and sphenoid bones. Large extensions of these sinuses may be found in the frontal bones above the orbits.

The ostia of the ethmoid sinuses are the smallest of any of the paranasal sinuses. The anterior ethmoid cells have smaller ostia than the posterior cells. The anterior group opens into the ethmoidal infundibulum or into the haitus semilunaris. The middle group opens into the middle meatus above the hiatus semilunaris, and the posterior group drains into the superior meatus.

The lateral wall of each ethmoid sinus is a thin bony plate, the lamina papyracea, which forms the medial orbital wall.

The midline between the ethmoid sinuses contains the superior portion of the nasal cavity. At the most superior portion of the nasal cavity lies the cribriform plate which connects the left and right groups of air sinuses. The roof of the ethmoid sinus forms a portion of the medial floor on the anterior cranial fossa lateral to the cribriform plate. The medial wall of each ethmoid sinus is formed by a thin lamella of bone from which arise the middle, superior and inferior turbinates. Anteriorly the attachment of the middle turbinate is at the lateral margin of the cribriform plate. The roof of the ethmoidal complex is formed by a medial extension of the orbital plate of the frontal bone which articulates with the cribriform plate, the fovea ethmoidalis.

The ethmoid cells can pneumatize the middle turbinate (concha bullosa) in 4–10% of cases.

The proximity of the posterior ethmoid cells to the orbital apex, optic canal and optic nerve can give rise to loss of vision as a complication of surgery on these sinuses.

Sphenoid sinus

The sphenoid sinus is a cavity behind the ethmoid cells that can extend posteriorly to the clivus. It is not visible at birth, but can be recognized radiologically at about 3 years of age. It enlarges progressively to reach adult size at the age of 10–12 years. The sphenoid sinuses are paired cavities divided by a septum. This is usually in the midline anteriorly but off-centre posteriorly, resulting in asymmetrical cavities. The sphenoid sinuses drain into the sphenoethmoidal recesses. The degree of pneumatization varies considerably. The average sinus measures $2 \times 2 \times 0.5$ cm in size. The posterior limit of the sinus is variable. Large sphenoid sinus extensions can occasionally be found in the skull base along the floor of the middle cranial fossa and in the superior portion of the pterygoid plates.

The roof of the sphenoid sinus is related anteriorly to the floor of the anterior cranial fossa, the optic chiasm and the sella turcica posteriorly. Lateral to the sphenoid sinus is the superior orbital fissure, the optic canal, optic nerve, cavernous sinus and the internal carotid artery. Posteriorly are situated the clivus, prepontine cistern, pons and basilar artery. The sinus floor is the roof the nasopharynx, and the anterior sinus wall is the back of the nasal fossa.

Maxillary antra

The maxillary sinus is the first of the sinuses to form. It appears within a few weeks of birth and enlarges throughout childhood. The developing maxillary sinus initially lies medial to the orbit. Its growth rate is approximately 2 mm vertically and 3 mm anteroposteriorly each year. In infancy the maxillary sinus floor lies at the level of the middle meatus. By the eighth to the ninth year the sinus floor is near the level of the nasal fossa floor. From this point there is considerable variation in the further growth of the inferior recess – it reaches the plane of the hard palate by the age of 12 years.

Grossly the maxillary sinuses tend to develop symmetrically. Unilateral and bilateral hypoplasia occur in 1.7% and 7.2% respectively.

The adult maxillary sinus is a large cavity, average volume 14.75 ml, lying in the body of the maxilla, triangular when viewed in the postero-anterior plane and quadrilateral when viewed in the sagittal plane. The highest point of the antrum is the posteromedial portion lying directly beneath the orbital apex. The roof is formed by the floor of the orbit and contains a ridge for the infraorbital canal. The floor of the antrum is formed by the alveolar process of the maxilla. This part of the antrum develops completely with the eruption of the permanent dentition. The medial wall of the maxillary antrum is formed by the lateral wall of the nasal cavity. Laterally, the maxillary antrum extends for a variable distance into the zygomatic bone, forming the zygomatic recess. Below this the lateral wall is relatively thin and often shows a defect due to the superior dental vessels.

The ostium of the sinus is on the highest part of the medial wall and is approximately 4 mm in diameter. The maxillary sinus drains into the ethmoidal infundibulum which, via the hiatus semilunaris, opens into the nasal cavity. The channel of the infundibulum is approximately 5 mm long and is directed upward and medially into the nasal fossa. The location of the ostium ensures that sinus drainage in the erect position is accomplished by cilial action. A narrow infundibulum can further interfere with sinus drainage.

Septa may occur within the antrum and may be partial or complete, the latter type being associated with separate ostia for the two chambers.

A posterior ethmoid air cell may at times enlarge and encroach on the medial wall of the maxillary antrum and may occupy a significant portion of its volume.

The medial wall of the antrum is surrounded by the air of the nasal cavity. Much of the remainder of the sinus is surrounded by fat density which is helpful in evaluating spread of disease beyond the bony limits of the maxillary sinus. Posterior to the maxillary sinus and anterior to the pterygoid plate lies the pterygopalatine fossa containing the terminal portion of the internal maxillary artery and the sphenopalatine ganglion.

In the adult skull the medial wall of the maxillary bone has a large hole, the maxillary hiatus, that exposes the interior of the meatus. This hole is covered in part by four bones. The perpendicular plate of the palatine bone lies posteriorly, while the lacrimal bone is situated anterosuperiorly. The inferior turbinate covers the inferior portion of the maxillary hiatus. Resting above the line of attachment of this turbinate is the uncinate process of the ethmoid bone. The remaining central portion of the maxillary hiatus is covered by the nasal and sinus mucous membranes in which are situated the infundibulum and maxillary ostium. This membranous area is important in efforts to irrigate the antrum.

Normal sectional anatomy is shown in Figs 5.1–5.3.

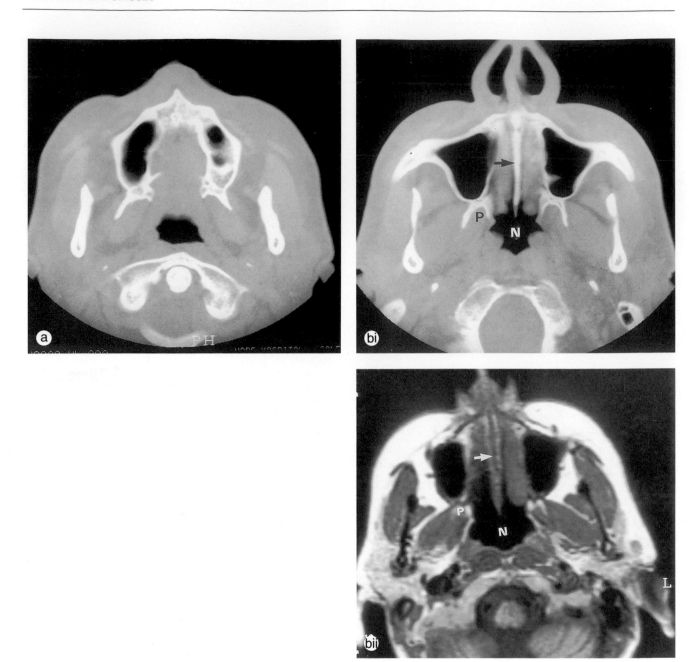

Fig. 5.1 Axial sectional anatomy: inferior to superior.
(a)   **CT.** The most inferior section shows the *maxillary alveolar ridge*. The inferior alveolar recesses of the maxillary sinuses can be seen pneumatizing the alveolar ridges and surrounding tooth roots. The soft palate is seen posterior to the hard palate. The pterygoid plates are well seen at this level.

(b)   (i) **CT**; (ii) **MRI – T1W.** *At the lowermost portion of the maxillary antra*, the lower nasal septum (arrow) is seen abutting the top of the hard palate. The pterygoid fossa, P, located between the medial and lateral pterygoid plates is seen, with the lowermost part of the nasopharynx, N, in the midline.

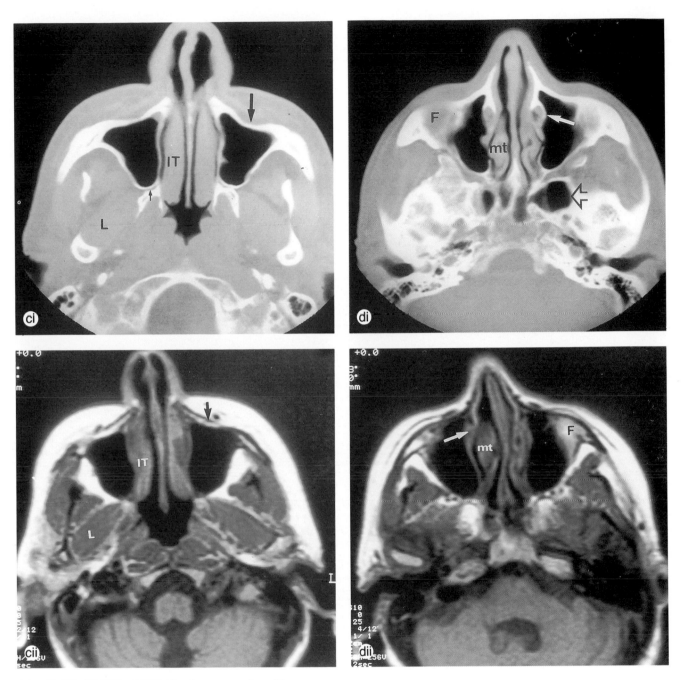

(c) (i) **CT**; (ii) **MRI – T1W.** *The lower nasal cavities* are seen with the inferior turbinates, IT. The inferior meati lie just lateral to each inferior turbinate. The medial walls of the maxillary sinuses are partially bone and partially membrane, the latter forming a C-shape which is closed posteriorly and bridges the posterior attachment to the inferior turbinate. The anterior wall of the antrum has a slightly concave anterior margin – the canine fossa (large arrow). The pterygopalatine fossa (small arrow) is seen just behind the medial posterior aspect of the antrum immediately anterior to the pterygoid process of the sphenoid bone. The sphenopalatine foramen connects the nasal fossa with the pterygopalatine fossa. The lateral pterygoid muscle, L, is seen extending from the pterygoid plate to the mandibular condyle.

(d) (i) **CT**; (ii) **MRI – T1W.** *At the level of the middle turbinate (mt) and middle meati.* The infraorbital foramen may create a focal indentation in the anterior antral wall. The nasolacrimal canal is in the medial antral wall anteriorly (arrow). *The inferior portion of the orbital floor* (F) lies antero-lateral in the orbit indenting the anterior wall of the antrum. Small lateral recesses of the sphenoid sinus (open arrow) in the greater wing of the sphenoid and into the anterior clinoid processes may be seen.

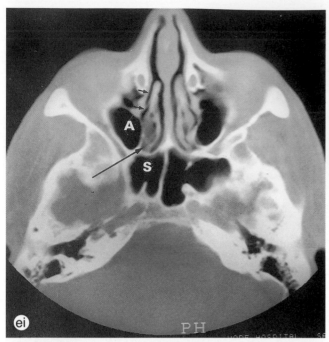

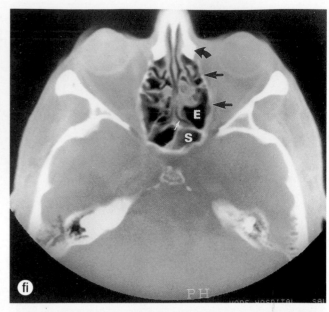

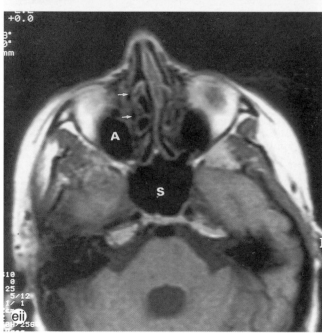

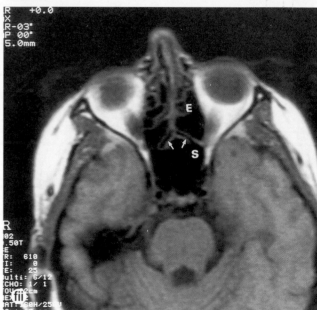

(e) (i) CT; (ii) MRI – T1W. The posterior lateral attachment of the middle turbinate is visible (long arrow). The middle meatus is just lateral to the turbinate (small arrow). The bases of the sphenoid sinus (S) are now seen. The superior recesses of the antra (A) are seen in the postero-medial orbital floor. These recesses appear as round or ovoid air spaces with a thin bony wall separating them from the orbital contents and should not be confused with the ethmoid air cells.

(f) (i) CT; (ii) MRI – T1W. *At the level of the ethmoid air cells* (E) the lamina papyracea (black arrows) forms the intermediate part of the medial orbital margin with the nasal and lacrimal bones anteriorly (curved arrow) and the lateral margin of the anterior sphenoid sinus (S) posteriorly. Note the position of the sphenoethmoidal recesses (small white arrow) into which the sphenoid sinus drains (see also Fig. 5.2(d)).

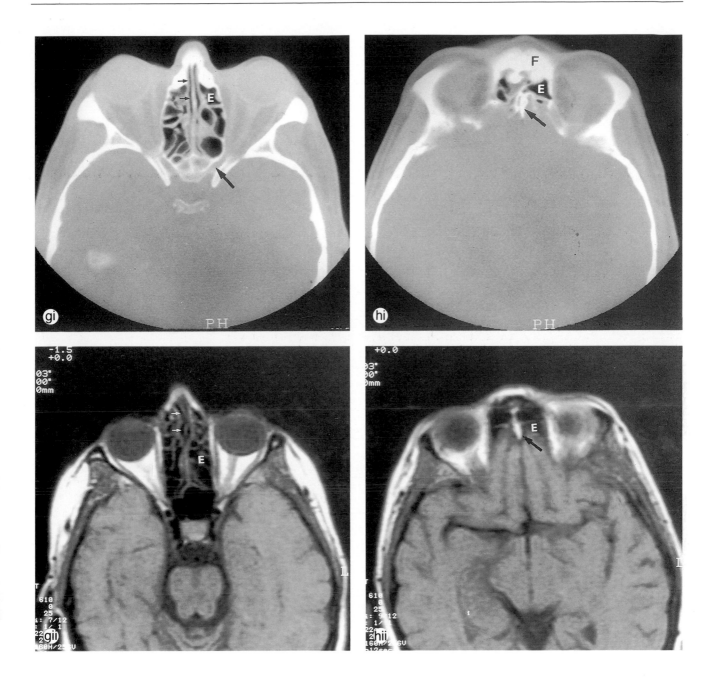

(g) **(i) CT**; **(ii) MRI – T1W.** *The olfactory recesses are seen as narrow slits (small arrows) on either side of the nasal septum with the ethmoid air sinuses (E) lateral to these recesses. The optic foramina are visible posteriorly (long arrows).*

(h) **(i) CT**; **(ii) MRI – T1W.** *At the level of the crista galli (arrow). Anteriorly are the bases of the frontal sinuses (F). The upper ethmoid sinuses (E) are seen posteriorly. The intracranial surface of the cribriform plates is identified by the crista galli.*

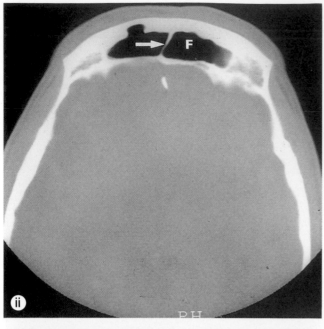

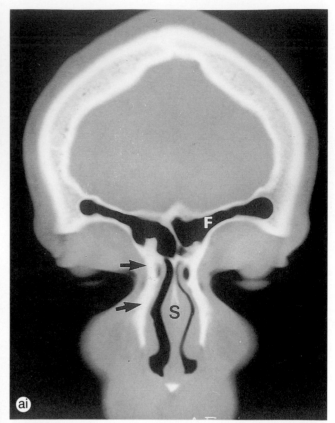

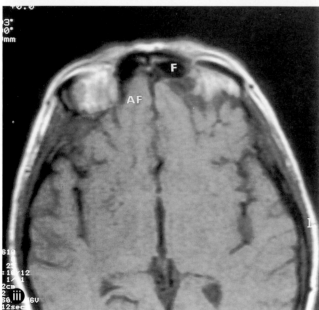

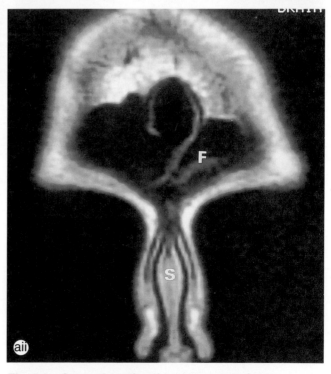

(i) (i) **CT**; (ii) **MRI – T1W.** *At the level of the frontal sinuses*
(F). The anterior and posterior walls of the sinus are well
seen. The intersinus septum (arrow) and small septations
and scalloped sinus margins are seen. The anterior cranial
fossa contents (AF) are seen.

**Fig. 5.2   Coronal sectional anatomy: anterior to
posterior.**
(a) (i) **CT**; (ii) **MRI – T1W.** *At the level of the frontal sinus*
(F). The intersinus septum and scalloped sinus margins are
seen. The frontal process of the maxilla (arrows), nasal
bones and nasal septum (S) are identified.

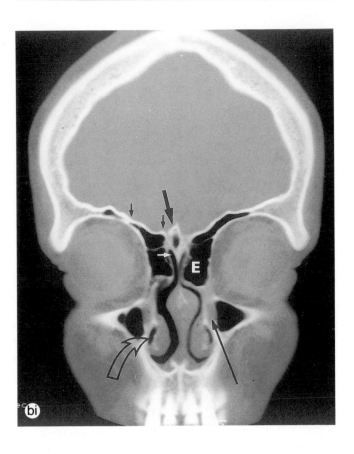

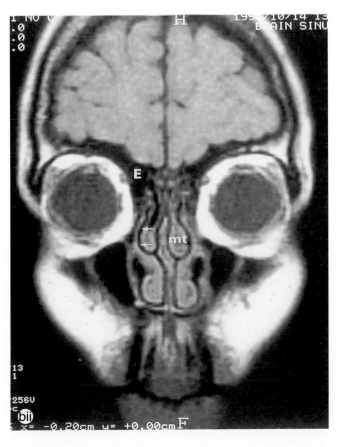

(b) (i) **CT**; (ii) **MRI – T1W.** *At the level of the crista galli* (large arrow). The fovea ethmoidales (small black arrows) are seen on either side. The nasal septum is prominent in the midline. The perpendicular plate of the ethmoid bone contributes to the nasal septum superiorly with the cartilaginous nasal septum caudally. The anterior ethmoid air cells (E) are visible. Supraorbital extension of the ethmoid air cells into the orbital roofs and a pneumatized crista galli may be visualized in the midline. The anterior portion of the maxillary sinus with the nasolacrimal duct entering its medial wall is identified (long arrow). The straight margin of the ethmoid lamina papyracea is seen forming the medial wall of the orbit. Both the superior and inferior orbital margins are now visualized. Superiorly, the olfactory recess of the nasal fossa (small white arrow in **i**) extends superiorly to the cribriform plate on each side of the crista galli. The middle turbinate (mt), attached anteriorly along the lateral margin of the cribriform plate is seen and localizes the roof of the nasal cavity. The anterior portion of the middle meatus (small white arrows in **ii**) is seen lateral to the middle concha. The anterior tips of the inferior conchae and the inferior meati (open arrow) are now visible.

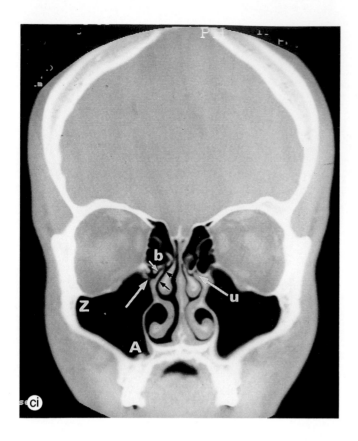

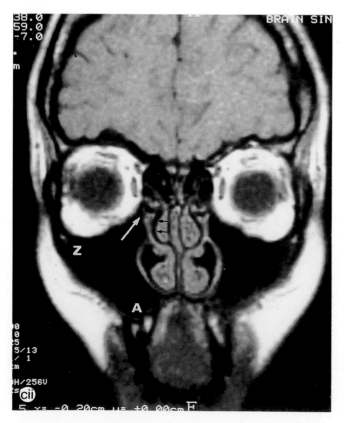

(c) **(i) CT**; **(ii) MRI – T1W.** *At the level of the ethmoid infundibulum and osteomeatal complex.* The infundibulum (small white arrow) lies above the uncinate process of the ethmoid bone and below the bulla ethmoidalis (b). The uncinate process (u) rests on the junction of the inferior concha with the medial wall of the maxilla. The maxillary antrum opens (long white arrow) into the lower lateral portion of the infundibulum, while the upper medial aspect of the infundibulum opens via the hiatus semilunaris (above the superior tip of the uncinate process) into the middle meatus (small black arrows). A portion of the medial antral wall in this region is formed entirely by mucosa. The main maxillary sinus cavity is seen with its zygomatic (Z) and alveolar (A) recesses. In the midline the cribriform plates lie at a lower level than the roofs of the ethmoid air cells. Note that detail of the osteomeatal complex is much better on CT than MR.

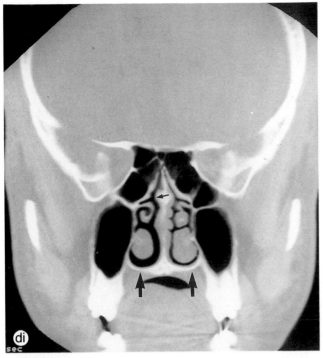

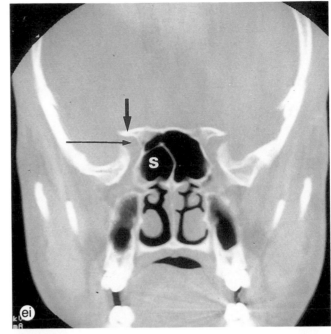

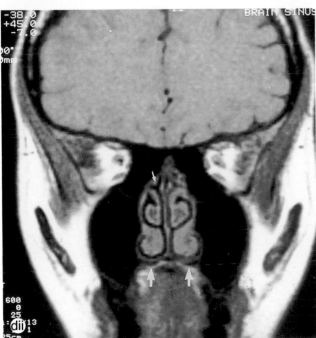

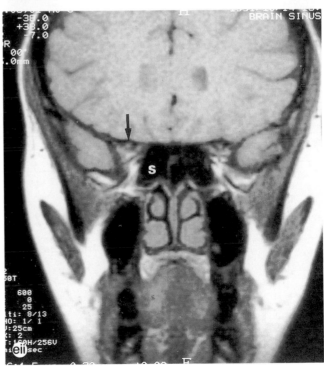

(d) (i) CT; (ii) MRI – T1W. *At the level of the superior turbinate*. The maxillary sinus now assumes an oval shape with its long axis in the craniocaudal plane. At this level the medial walls of the antra are formed by the vertical plate of the palatine bone. Inferiorly the hard palate (long arrows) is formed by the horizontal plates of the palatine bone. The olfactory recess on either side of the nasal septum widens into the sphenoethmoidal recess (small arrows) into which the ostia of the sphenoid sinus open. This marks the junction between the ethmoid and sphenoid bones and lies posterior to the superior meatus.

(e) (i) CT; (ii) MRI – T1W. *At the level of the anterior portions of the sphenoid sinus (S)*. The antra are more elongated and appear smaller. The optic canal (long arrow) is lateral to the sphenoid sinus. The anterior clinoid process (short arrow) is just above the optic canal.

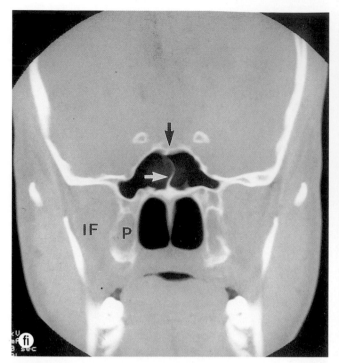

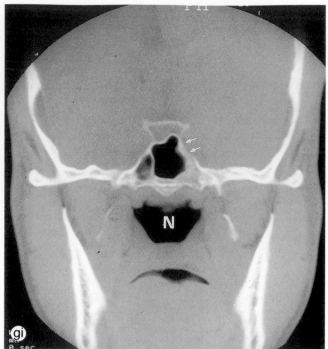

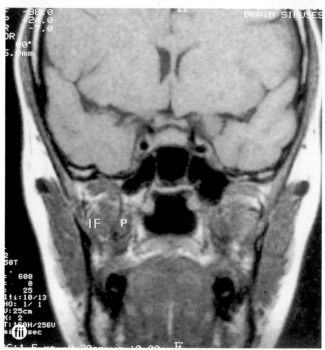

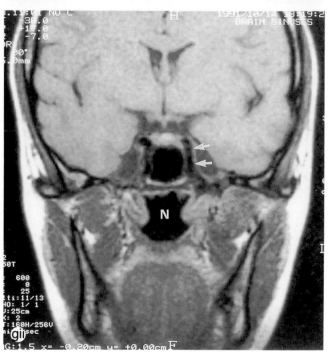

(f) **(i) CT; (ii) MRI – T1W.** *At the level of the pterygoid plates.* The interspinus septum (white arrow) of the sphenoid sinus is in the midline anteriorly, but posteriorly may be angulated sharply to one side, creating two unequally sized sinuses. The medial pterygoid plate is almost vertically orientated. The lateral plate is tilted so that its lower end is lateral. The pterygoid fossa (P) lies between these plates with the infratemporal fossa (IF) lying laterally. The optic canal may cause an indentation on the upper lateral sinus wall – this relates sphenoid sinus disease to orbital symptoms. The roof of the sphenoid sinus is formed by the planum sphenoidale (black arrow).

(g) **(i) CT; (ii) MRI – T1W.** *At the level of the cavernous sinus.* In (i), the carotid impression (arrows) is seen on the upper, lateral sphenoid sinus. The roof of the sphenoid sinus is now the floor of the sella turcica. Beneath the sinus floor is the roof of the nasopharynx. In (ii), the flow void of the internal carotid artery itself is seen (arrows).

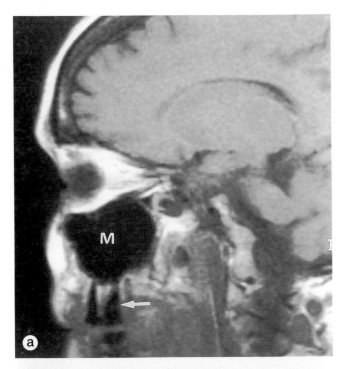

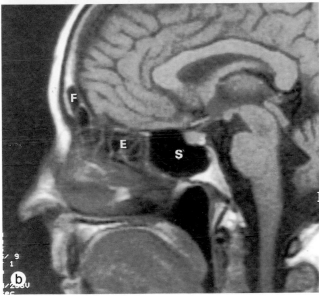

Fig. 5.3 Sagittal sectional anatomy: paramedian to median.

(a) **MRI – T1W.** *Through the mid-portion of the maxillary antrum (m)* the configuration of the antrum is rectangular. Behind the antrum, the medial aspect of the infratemporal fossa narrows into the sphenomaxillary fissure. The medial and lateral pterygoid muscles are behind the sinus. The largest portion of the antrum is its most medial part. The maxillary teeth in the maxillary alveolus are seen directly below the sinus (arrow).

(b) **MRI – T1W.** *The frontal sinuses (F), ethmoid air cells (E) and sphenoid (S) sinus* are clearly seen in this midline section. The sphenoid sinus is best seen in the sagittal plane. The roof of the sphenoid sinus (planum sphenoidale) lies directly behind the cribriform plate.

Abnormalities of development

The paranasal sinuses show a great variation in their degree of development both unilaterally and bilaterally. The frontal sinuses are often under-developed or absent and the sphenoid sinus is frequently small. The maxillary antra and the ethmoid air cells show less variation in size.

Under-developed sinuses are often associated with hypoplastic conditions of the facial skeleton and skull base including cleidocranial dysostosis, craniofacial dysostosis, achondroplasia and pycno-dysostosis. Hypoplasia or absence of the frontal sinuses is common in Down's syndrome. In cranio-stenosis there is unilateral hypoplasia of the skull with under-development of the frontal sinus on the affected side. Arrested development of the maxillary sinuses may occur as a result of antral infection in infancy or childhood.

Excess pneumatization with over-development of the sinuses is commonly found in acromegaly and in cases of arrested hydrocephalus. Unilateral over-development of the paranasal sinuses, particularly the frontal and/or ethmoid sinuses, may be seen in congenital cerebral hemiatrophy.

INFLAMMATORY CONDITIONS

Headache, facial pain and swelling and nasal discharge are symptoms of infective sinusitis. Acute bacterial sinusitis is usually due to ostial obstruction and there is more often unilateral or isolated sinus involvement. Acute sinusitis is most often due to secondary bacterial infection following an upper respiratory tract infection of viral origin. Infection can also occur in the maxillary antrum by secondary extension from an infected tooth in the upper jaw. The primary site of infection is the lining mucosa of the sinuses.

Chronic sinusitis may result as a complication of acute sinusitis or persistent acute inflammation. It is generally manifest as a thickening of the lining mucosa of the sinus. The mucosa may become either hypertrophied and polypoid or atrophic and sclerosed. Secondary reactive changes may occur in the bony walls of the chronically infected sinuses which become thickened and sclerotic. In some cases demineralization occurs. The frontal sinuses are most commonly affected in chronic sinusitis. The condition may be associated with recurrent orbital cellulitis and, rarely, pseudotumour or chronic inflammation in the orbit.

113

Imaging

Normal paranasal sinus mucosa is so thin that it cannot be visualized on imaging, including plain radiographs. Inflammatory changes in the paranasal sinuses result in diffuse thickening of the mucosa and submucosa which can be visualized on plain radiographs, CT scanning and MRI. Mucosal thickening of the walls of the paranasal sinuses can vary from 1–2 mm to polypoid hypertrophy which may fill the sinus. Thickened mucosa appears as a uniform soft-tissue density zone that separates the sinus air from the bone. On plain films this mucosal thickening is more readily recognized in the frontal sinuses and along the lateral walls of the maxillary antra. On post-contrast CT scans mucosal thickening due to active infection shows surface enhancement with submucosal oedema which does not enhance.

However, asymptomatic infection is common, being revealed by the presence of clouding or mucosal thickening of the sinuses. The highest incidence occurs in the ethmoid air cells (10.9–28%) [2,3]. Isolated ethmoid clouding on CT scanning was observed in 15% and is likely to be found in one in seven of the adult population.

The mucosal thickening of infective sinusitis shows contrast enhancement on CT or MRI (Fig. 5.4) in contradistinction to the mucosa thickened by allergy or fibrosis which does not enhance on post-contrast CT and MRI. Moderate mucosal thickening results in a uniformly thick soft-tissue density on sectional imaging (Figs 5.5 and 5.6). However, more redundant mucosa can appear almost nodular or 'wave like' (Fig. 5.7).

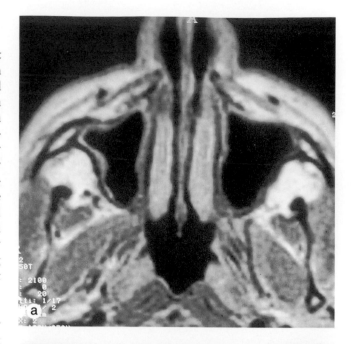

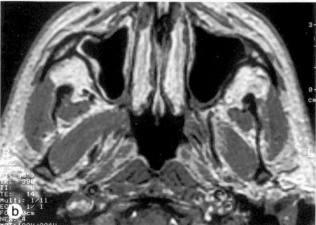

Fig. 5.4 Infective mucosal thickening: T1W axial section. (**a**) Pre-contrast; (**b**) Following i.v. gadolinium DTPA. The mucosal thickening is shown as a smooth, uniform line of soft tissue surrounding the walls of the antra and demonstrates moderate contrast enhancement.

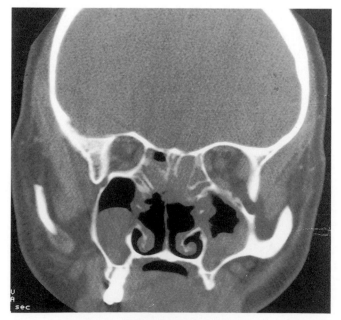

Fig. 5.5 Moderate mucosal thickening: maxillary antra and ethmoid sinuses. Coronal CT (no contrast). Uniform, polypoid mucosal thickening in both antra and ethmoid sinuses producing a lobulated appearance in the left antrum.

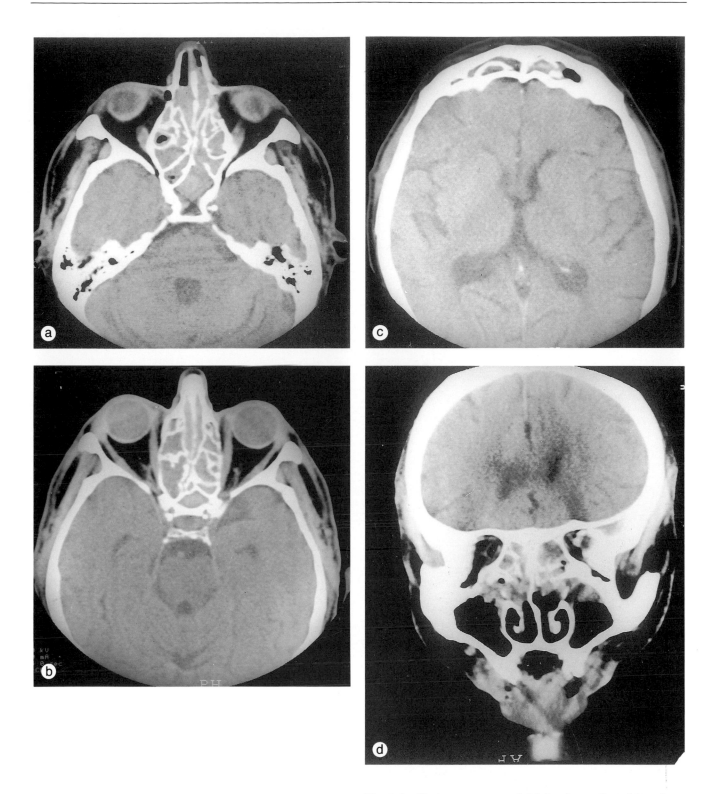

Fig. 5.6 Moderate mucosal thickening: ethmoid and frontal sinuses. Plain CT in axial (**a**, **b**, **c**) and coronal (**d**) planes. The posterior, middle and anterior ethmoid sinuses are opaque, showing diffuse soft-tissue density and widening of the posterior and middle ethmoid complexes with extension into the nasal cavity and sphenoid sinus (**a**), frontal sinus (**c**), and postero-superior portions of the antra (**d**).

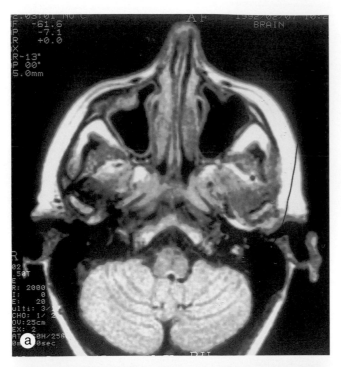

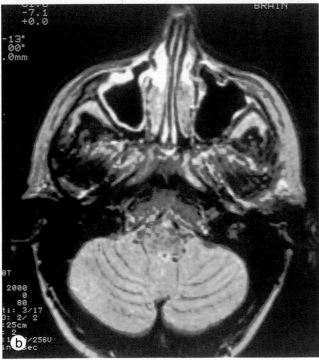

Fig. 5.7 Mucosal thickening: maxillary antra. MRI axial images (**a**) T1W and (**b**) T2W. There is smooth, uniform mucosal thickening in both antra shown as (**a**) intermediate on the T1W image and (**b**) increased signal on the T2W image, typical of high water content.

images, intermediate intensity on proton-density-weighted images and high signal on T2-weighted images. The long T2-weighted signal is helpful in differentiating inflamed tissue from tumours, which have an intermediate signal on T2-weighted images. Fibrosis, however, also has an intermediate signal intensity on all sequences and can therefore be differentiated from actively inflamed tissue on T2-weighted sequences. However, MRI cannot reliably differentiate fibrosis from tumour within the sinus cavity. Furthermore, in the case of chronic infections, the secretions may become desiccated [4]. These secretions lose some of their signal intensity and display intermediate signal intensity on both T1- and T2-weighted images. Thus an opacified sinus due to chronic infection seen on CT may also display intermediate signal characteristics on both T1- and T2-weighted images, precluding confident exclusion of a tumour. Contrast-enhanced MRI is required in these cases. In inflammation a peripheral rim of enhancement is usual, whereas in the case of a tumour there is a solid or homogeneously enhancing sinus cavity.

Acute sinusitis

An air–fluid level is the cardinal sign and is most frequently seen in acute bacterial sinusitis, best seen in the maxillary antrum (Fig. 5.8). Sinus lavage following treatment of acute bacterial sinusitis is the next most common cause. Trauma is also a frequent cause of an air–fluid level whether or not there is an associated fracture of the wall of the sinus. The fluid in the sinus may be due to a mucosal tear. On CT scanning, blood in the sinus has a higher attenuation than mucosal oedema and inflammatory secretion. On MRI fresh blood has a low signal intensity on T1-weighted and proton-density-weighted images and a high signal after 24–48 hours. On the other hand, acute inflammatory tissue demonstrates a low to intermediate signal on T1-weighted and proton-density-weighted images. Haemorrhage resulting in fluid levels in the sinus cavity may also result from barotrauma and bleeding disorders such as von Willebrand's disease, in which bleeding tends to occur at mucosal surfaces.

Air–fluid levels are not usually seen in allergic cases.

Where the secretions are retained, multiple concentric rings appear in the sinus on CT. These rings represent layers formed by the outer bony margin, a low density submucosal layer and a thin enhancing mucosal ring with a central zone of low density in the cavity of the sinus due to retained secretions.

On MRI, mucosal thickening secondary to inflammation or allergy has signal characteristics typical of high water content; low signal on T1-weighted

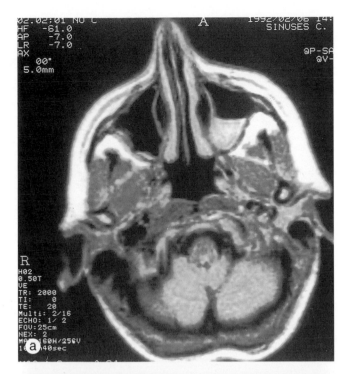

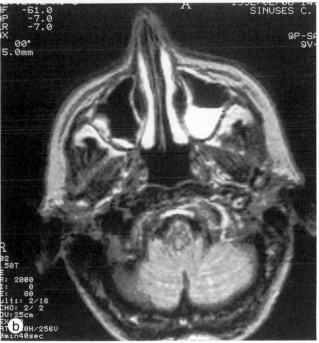

Fig. 5.8 Acute sinusitis. MRI axial images (**a**) T1W and (**b**) T2W. An air–fluid level is seen in the left maxillary antrum which shows intermediate signal on the T1W and high signal on the T2W image, typical of high water content. Note mucosal thickening of the right antrum and nasal mucosa.

An air–fluid level in the maxillary antrum usually indicates acute bacterial sinusitis or recent antral lavage and less commonly trauma or obstruction caused by a nasogastric tube or mass.

Allergy

Allergy may produce a spectrum of changes in the paranasal sinuses varying from mild mucosal thickening to complete opacification of the sinus [5]. The imaging features may be difficult to distinguish from those of infective sinusitis, especially as the two conditions may coexist. Thickening of the nasal turbinates is characteristic of allergic sinusitis and is usually accompanied by the polypid type of mucosal thickening in the adjacent sinuses, the thickened mucosa producing a convex indentation into the cavity of the sinus. The imaging characteristics reflect the high water content of the mucosa, with non-enhancement on post-contrast CT scans as opposed to acute bacterial infections which show surface enhancement. The condition is frequently complicated by nasal polyposis. Air–fluid levels in the sinus cavities are uncommon and when present usually indicate superadded infection.

Mucous retention cysts

Retention cysts are seen frequently on imaging. The most common are mucous retention cysts which are lined by epithelium of an obstructed seromucinous gland. Mucous retention cysts may be found on plain films in about 10% of patients and occur most frequently in the maxillary sinus, but also occur in the frontal and ethmoid sinuses.

Serous cysts result from fluid accumulation in the submucosal layer of the sinus [5] and tend to occur in the base of the maxillary sinuses.

Polyps

Inflammatory polyps are indistinguishable from allergic polyps on imaging (Fig. 5.9 and 5.10). Polyps are associated with asthma and with reactions to drugs, particularly aspirin and nickel. The presence

Significance of an air–fluid level

In the frontal sinuses, an air–fluid level indicates acute bacterial sinusitis. In the sphenoid sinus it indicates the presence of acute sinusitis, nasal cavity obstruction or haemorrhage or cerebrospinal fluid from a fracture of the skull base. These fractures usually involve the floor of the anterior cranial fossa or the mastoid portion of the temporal bone rather than the sphenoid sinus and cause CSF rhinorrhoea.

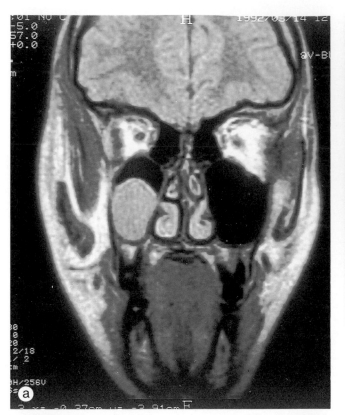

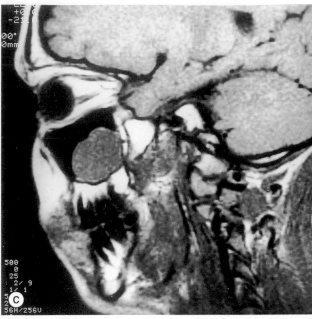

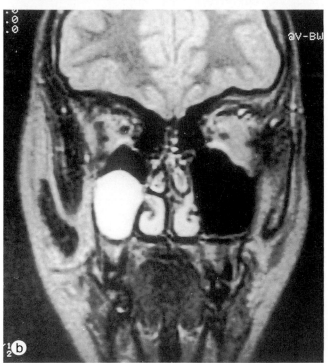

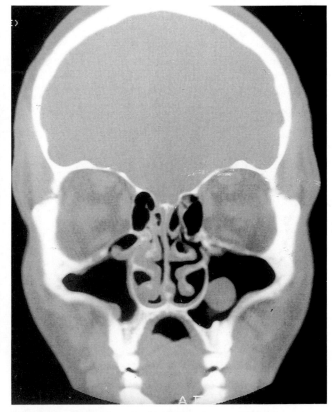

Fig. 5.9   Large polyp in right antrum. MRI images in coronal and sagittal planes. (**a**) Coronal T1W; (**b**) coronal T2W; (**c**) sagittal T1W. Large polyp in the postero-inferior aspect of the right antrum showing smooth margins with a convex superior margin. Characteristic intermediate signal on T1W and high signal on T2W, indicating high water content.

Fig. 5.10   Multiple antral polyps. A coronal CT (no contrast) shows well-defined masses displaying homogeneous soft-tissue density in both antra. Note moderate mucosal thickening of the right nasal cavity.

of a polyp in a child should suggest possible cystic fibrosis. Polyps may occur secondary to infectious rhinosinusitis and there is an association between nasal polyps and diabetes mellitus.

Polyps result from a local heaping-up of the mucosa. Accumulation of intercellular fluid results in an increase in polyp size. The retention cyst and polyp cannot be differentiated on imaging. On plain films and CT these lesions display homogeneous soft-tissue characteristics with smooth, convex borders. The lesions may be multiple or single and most are small, not filling the sinus cavity. A cyst in the antral roof may simulate a blow-out fracture, requiring coronal CT scanning for differentiation.

The surface of a large cyst in the antrum may simulate an air–fluid level. The differentiation is made on axial imaging which demonstrates the convex polypoid margins. MRI displays typical signal characteristics of paranasal secretions, i.e. high water and low specific protein content, giving low to intermediate signal on T1- or protein-density-weighted images and high signal on T2-weighted images. Increased protein or infection occurring within the cyst produces a shortening of the T1 relaxation time, resulting in a high signal intensity. Chronic polyps eroding through the floor of the anterior cranial fossa have different MRI characteristics. The polyps show an increased protein content and resorption of their water content [4]. This results in shortening of the T1 and T2 relaxation times with non-homogeneity in all sequences which differentiate these polyps from tumours. Very proteinaceous polyps may demonstrate signal voids.

Polyps are the most common expansile lesions in the nasal cavity. Multiple nasal polyps may present as a mass in the nasal cavity that may be difficult to distinguish from a tumour (Figs 5.11 and 5.12).

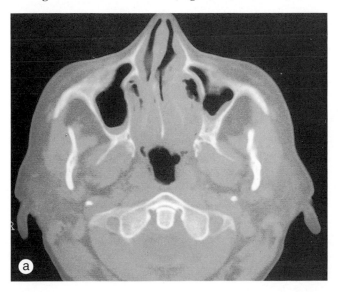

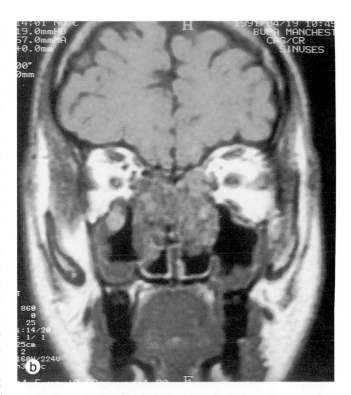

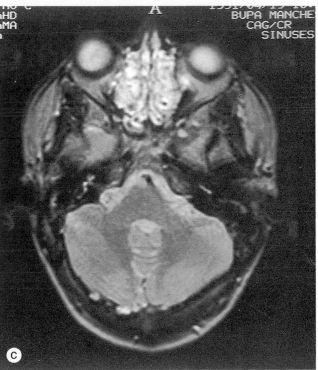

Fig. 5.11 Nasal polyps. CT and MRI (**a**) Axial CT scan shows large soft-tissue masses in the nasal cavities extending to the postnasal space. Note mucosal thickening of the antra. (**b**) Coronal MR T1W and (**c**) axial T2W images. The polyps show typical intermediate signal on T1W and high signal on T2W images, indicating high water content. Note mucosal thickening of both maxillary antra.

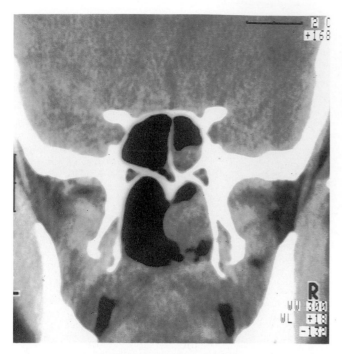

Fig. 5.12 Large polyp in postnasal space. Coronal CT (no contrast). Well-defined soft-tissue density mass in the left side of the postnasal space. Note mucosal thickening of the left compartment of the sphenoid sinus.

Multiple polyps in the ethmoid sinuses can be differentiated from tumour by demonstrating intact bony septa [6,7] (Fig. 5.13). The demonstration of a thin rim of mucoid material separating the polypoid masses from the adjacent bones is the differentiating feature. In tumours the soft-tissue abuts the bone directly. The demonstration of this thin hypodense rim on CT scanning indicates the presence of a mass of polyps.

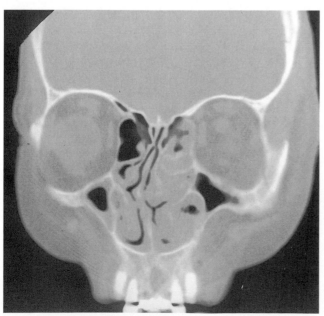

The hypodense rim may also be caused by myce-tomas, following intrasinus haemorrhage and occasionally in association with inspissated secretions and surrounding oedema [8]. Thus although this sign is non-specific its presence reliably indicates that a soft-tissue lesion is not a malignant tumour [9].

Antrochoanal polyp

This polyp originates from the maxillary sinus and extends through the ostium to the nasal cavity and nasopharynx. These solitary, unilateral lesions account for 4–6% of all nasal polyps. Most occur in young adults and should be treated via a Caldwell–Luc approach to prevent recurrence due to incomplete removal. A small antrochoanal polyp completely fills the antrum, producing widening of the infundibular region and slight extrusion of the mass into the lateral wall of the nasal fossa. As the polyp enlarges it destroys the medial wall of the antrum, filling the ipsilateral nasal cavity, and extends posteriorly into the nasopharynx. Sectional imaging in the coronal plane is best for demonstrating the size and location of such polyps (Figs 5.14 and 5.15).

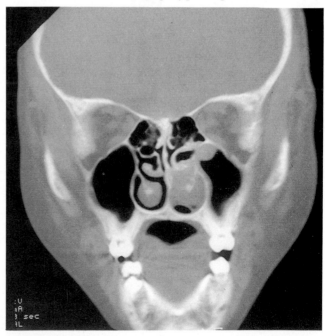

Fig. 5.14 Antro-choanal polyp. Coronal CT (no contrast) showing smooth polyp expanding the nasal cavity and extending into the left maxillary antrum.

Fig. 5.13 Multiple polyps: ethmoid sinus and nasal cavity. Coronal CT (no contrast). Multiple soft-tissue density polyps in the left ethmoid complex producing expansion of the loculi extending into the nasal cavities, particularly on the left. Note displacement of the nasal septum.

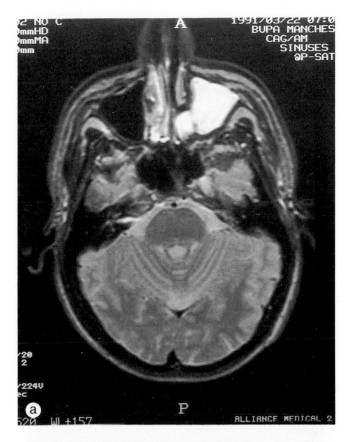

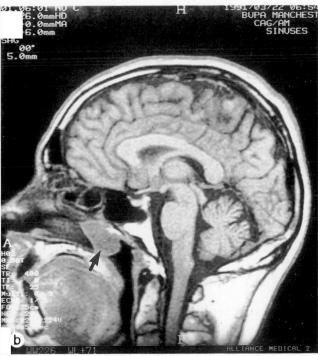

Fig. 5.15 Antro-choanal polyp. MRI (**a**) Axial T2W;
(**b**) Sagittal T1W. Typical high-signal polyp, indicating high
water content, in the left maxillary antrum prolapsing into
the nasal cavity. The polyp is seen as intermediate signal
on the T1W sagittal image in the posterior nasal cavity
(arrow).

Mucocele

This is the most common expansile lesion of the
paranasal sinuses. A mucocele is an airless, expanded sinus cavity that develops after obstruction
of the ostium and contains mucoid secretions produced by the mucosa. The obstruction may be
caused by infection, trauma or an underlying
tumour. This results in thinning and remodelling of
the sinus wall. Mucoceles are seen most often in the
frontal sinus (60–65%) and less often in the ethmoid
sinus (20–25%) [10], sphenoid sinus (1–2%) [11] or
maxillary sinuses (10%).

Mucoceles present with signs and symptoms of
the mass itself – proptosis, bossing of the forehead,
a mass in the superomedial orbit or nasal obstruction. Pain is rare and indicates an infected mucocele
or pyocele.

Imaging characteristics of mucoceles

Mucoceles cause opacification of the involved sinus
with expansion of the bony margins and loss of the
normal sinus scalloping. In the frontal sinus, there is
loss of bone density with loss of the normal dense
outline of the mucoperiosteum. The sinus cavity
presents a smooth contour with an ovoid or rounded
appearance. The erosion results in change of shape
and density of the superomedial margin of the orbit
which is readily detected. Downward and outward
displacement of the superomedial orbital rim is
almost pathognomonic of a frontal sinus mucocele.
Plain radiographs can underestimate the extent of
bony destruction by not demonstrating the erosion
of the lateral compartments. In addition to this
vertical extension mucoceles can extend posteriorly
into the orbital roof and subsequently cranially into
the floor of the anterior cranial fossa and caudally
into the orbital roof.

A mucocele of the ethmoid sinus usually originates in the anterior ethmoid air cells due to obstruc-
tions of the smaller ostia and less frequently in the
supraorbital air cells. It presents as an expansile
lesion that thins and remodels the lamina papyracea,
bowing it into the orbit and causing lateral displacement of the globe. This is frequently difficult to
demonstrate on plain radiographs. The presence of
an air–fluid level in the ethmoid air cells usually
indicates the presence of a mucocele that has ruptured, draining some of its contents into the nasal
cavity and adjacent ethmoid air cells.

Mucoceles of the maxillary antra result in complete opacification of the sinus cavity with expansion
of the walls. An advanced mucocele may produce
destruction of the antral wall making differentiation
from a carcinoma impossible.

Sphenoid sinus mucoceles expand anterolaterally into the posterior ethmoid air cells and the orbital apex. Less commonly, intracranial extension occurs into the sella turcica and cavernous sinus or downward into the nasopharynx. Large mucoceles may cause optic nerve compression. Imaging must demonstrate the relationship of the mass to the optic nerve (Fig. 5.16). On plain radiographs sphenoid mucoceles are liable to be misdiagnosed as tumours or as nasopharyngeal tumours invading the sphenoid. Typical conventional radiographic features are elevation or bone destruction in the floor of the pituitary fossa on the lateral projection. Progressive expansion will also cause erosion of the medial wall of the optic canal and elevation of the planum sphenoidale.

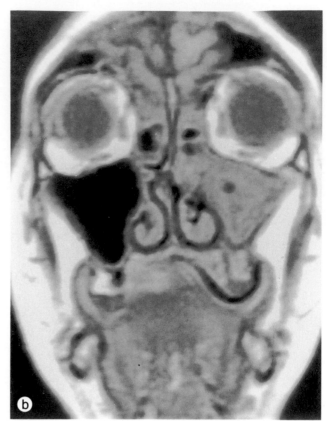

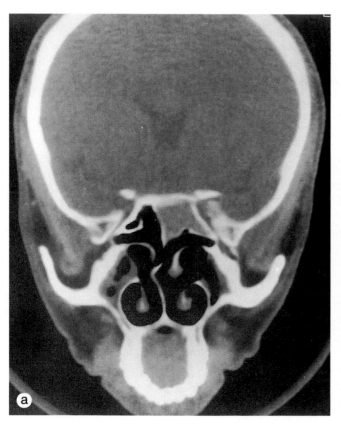

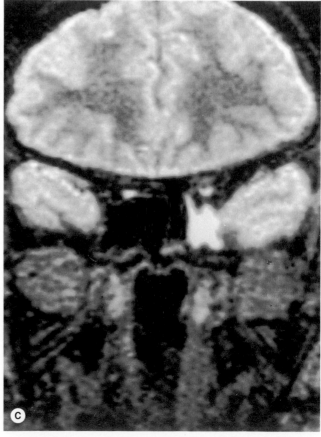

Fig. 5.16 Maxillary and sphenoid sinusitis. CT and MRI. (**a**) Coronal CT (no contrast); (**b**) coronal MRI T1W; (**c**) coronal MRI T2W. Note typical changes of sinusitis with intermediate signal on T1W and increased signal on T2W images in sinus cavities. The CT scan shows opaque left sphenoid sinus. Patient had symptoms of optic neuritis which resolved on treatment of sinusitis. Note absence of orbital involvement. (Reproduced by courtesy of Dr A Jackson.)

Sectional imaging in mucoceles

CT scanning in the axial and coronal planes demonstrates the expansion of the sinus cavity, bone remodelling, bone destruction and the soft-tissue mass. The mucocele is demonstrated as a homogeneous substance of mucoid attenuation (10–18 HU). In chronic mucoceles with a high protein content, the attenuation is higher (20–40 HU).

The lining mucosa may show a variable degree of enhancement following intravenous contrast administration. Contrast enhancement is necessary for showing involvement of the cavernous sinuses and intraorbital and intracranial extension.

Axial sections are necessary for demonstrating the expansion of the anterior and posterior walls in frontal mucoceles and for demonstrating the medial walls of the ethmoid air cells. Antral mucoceles are well shown on axial sections, particularly demonstrating any extension into the nasal fossa. The partial volume effect of the thinned walls may simulate destruction by the sinus cavity mucocele.

Coronal sections are best for demonstrating superior and inferior extension in the case of frontal, ethmoid and sphenoid mucoceles. Orbital extension is well demonstrated in this plane which should also be employed for demonstrating cavernous sinus involvement and intracranial extension (Figs 5.17–5.19).

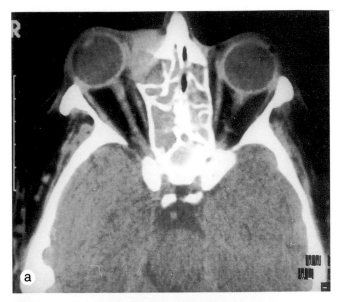

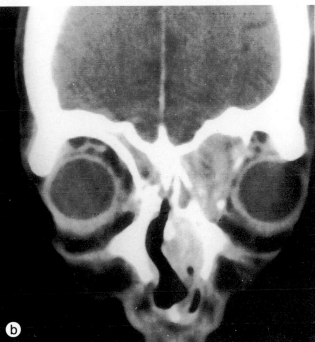

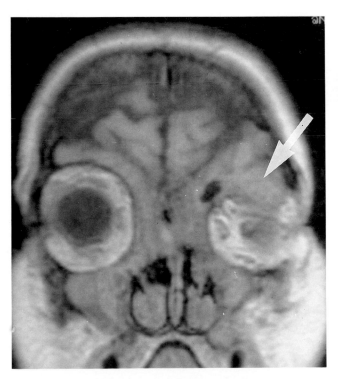

Fig. 5.17 Ethmoid mucocele. CT scans (no contrast). (**a**) Axial section; (**b**) coronal section. The ethmoid air sinuses bilaterally are opaque with expansile destruction of the right anterior compartment. The mucocele is seen as a mass of soft-tissue density in the antero-medial aspect of the right orbit with forward and lateral displacement of the globe.

Fig. 5.18 Fronto-ethmoidal mucocele. MRI. Coronal T1W image. Intermediate signal mass in right compartment of the frontal sinus (arrow) extending into the superior aspect of the right orbit, displacing the globe. Note pansinusitis. (Reproduced by courtesy of Dr A Jackson.)

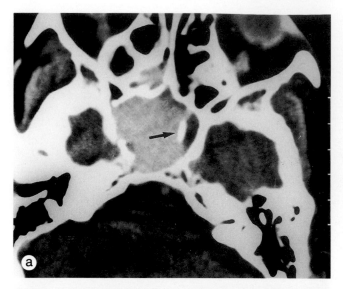

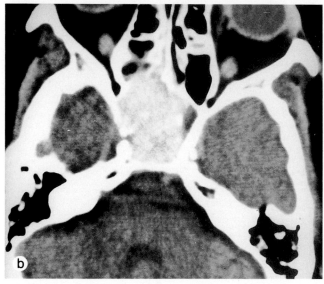

Fig. 5.19 Sphenoid mucocele. CT scan. Postcontrast axial sections. Enhancing soft-tissue density expanding the right compartment of the sphenoid sinus, extending anteriorly into the posterior ethmoid air cells. The intersinus bony septum is partially eroded (arrow). Note mucosal thickening of the ethmoid sinuses.

On MRI the signal characteristics are those of retained mucoid secretions in the sinuses: low to intermediate on T1-weighted and high on T2-weighted images. However, extreme differences are demonstrated [12] and reflect the variable hydration of the contents. The T2-weighted signal may be decreased and the T1-weighted signal higher due either to inspissated mucin or the presence of crystalline structures in patients with superimposed fungal infections, e.g. aspergillosis. The T1-weighted relaxation time shortening probably can be used as an indication of how long the mucocele has been present. Residual air in the sinus is seen as a signal void.

Thus mucoceles can have the following progressive MRI appearances; low T1W, high T2W; intermediate T1W, high T2W; high T1W, high T2W; intermediate to high T1W, low T2W; low T1W, low T2W [4]. If the signal characteristics are typical of well-hydrated secretions, a diagnosis of mucocele may be made with confidence. When the signal characteristics are intermediate on both T1W and T2W images, enhanced MRI may be useful for differentiating mucoceles from tumours [13].

A frontal meningocele presenting as a well-defined midline bone defect with absence or hypoplasia of the sinus, may be confused with mucocele of the frontal sinus.

Osteomyelitis

Osteomyelitis may occur as a complication of chronic sinusitis or fungal disease, and may also be seen as a complication of irradiation. The radiological features include focal bone rarefaction, sequestrum formation, reactive bone thickening, bony sclerosis and ultimately bone destruction and fragmentation.

Intracranial complications

These are infrequent and include meningitis, epidural abscess, subdural abscess, cerebritis and intracerebral abscess [14]. Only 3% of intracranial abscesses arise from paranasal sinus cavity disease.

Orbital complications

Orbital cellulitis and cavernous sinus thrombosis [15] occur as infrequent complications and are discussed in Chapter 4.

Fungal infections

The paranasal sinuses may be involved in aspergillosis, mucormyocosis, candidiasis, histoplasmosis, cryptococcosis and coccidiomycosis. The imaging characteristics span a wide spectrum. In the early stages, a non-specific mucosal inflammation may be present, mainly in the maxillary sinus or ethmoid air cells. Aspergilloma may affect the sphenoid sinus. The frontal sinuses are only rarely involved. Air–fluid levels are very uncommon and suggest bacterial infection. In fungal infection the surrounding bone may be thickened and sclerotic or eroded and remodelled [16]. The associated soft-tissue mass may be present in both the paranasal sinuses and nasal cavity and extend into the cheek, unlike in bacterial sinusitis. Aspergillosis commonly involves the lungs. However sinus aspergillosis is rare, but is the commonest cause of fungal disease. The incidence of sinonasal fungal infection has increased because of the increasing use of antibiotics, steroids, cytotoxics and other chemotherapeutic agents [17].

There are two types of sinus aspergillosis: invasive and non-invasive. The main difference between the two is the bone destruction that typifies the invasive type. The maxillary sinus is the most common site and may be affected either unilaterally or bilaterally [18], followed by the ethmoid sinus and the nasal cavity.

CT scanning is the method of choice for demonstrating the presence of calcified concretions, bone erosion and extension into the surrounding tissues. The most characteristic pattern is a multiple, linear interlacing network of high density [9]. Concentric and polypoidal mucosal thickening may also be present. Slightly enhancing mucosal masses filling the sinus cavity, separated from the sinus wall by a thin rim of mucoid density material which has a lower attenuation value than the fungal mass, may be demonstrated [18]. The commonest intracranial complication of sinonasal aspergillosis is granuloma formation which is in continuity with the sinus aspergillosis.

The differential diagnosis of sinus aspergillosis includes chronic bacterial sinusitis, malignant and benign neoplasms, tuberculosis, osteomyelitis, Wegener's granuloma, rhinoscleroma and mucormycosis, polyps and fibrous dysplasia [17].

Mucormycosis is more prevalent in poorly controlled diabetics and in patients with haematological malignancies (e.g. acute leukaemia), chronic renal failure, malnutrition, cancer and cirrhosis and in those undergoing chemotherapy. The organism is highly invasive and tends to spread rapidly from the nasal cavities to the paranasal sinuses (Fig. 5.20).

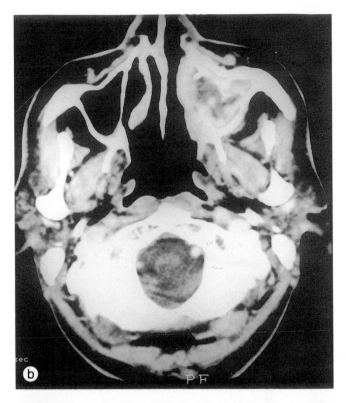

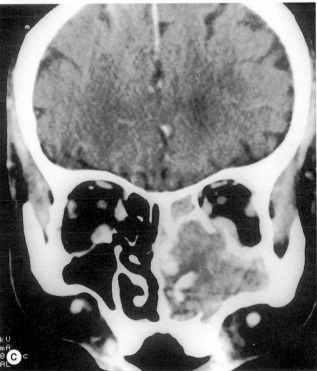

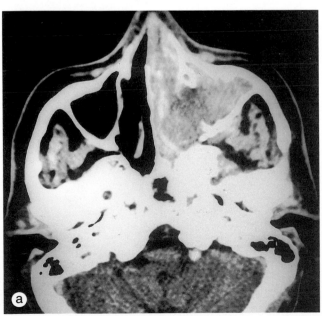

Fig. 5.20 Mucormycosis. Axial and coronal CT scans. Axial planes (**a**) precontrast and (**b**) postcontrast and (**c**) coronal plane postcontrast. Non-specific irregular enhancing mucosal mass in the right maxillary antrum with a central area of non-enhancement. Note extensive destruction of the medial and lateral walls of the antrum with extension of the mass into the right nasal cavity.

From the sinuses there is invasion of the orbits and cavernous sinuses via the ophthalmic vessels. The infection invades blood vessels, causing endothelial damage that initiates thrombosis, ischaemic and haemorrhagic infarction and finally purulent inflammation.

The MRI features of the mycetomas associated with fungal paranasal sinus disease may produce appearances difficult to interpret correctly, with signal voids seen on both T1- and T2-weighted sequences. This appearance reflects the thick, cheesy composition of the desiccated mycelial mass and possibly the production of haemosiderin, iron and manganese. A completely opacified sinus on CT may therefore look partially or totally aerated on MRI. Similar MRI findings can be encountered in chronic, non-fungal sinus disease, where highly proteinaceous and dessicated sinus secretions may also result in a signal void on T1- and T2-weighted images [4, 19].

GRANULOMA-PRODUCING DISEASES

The granulomatous diseases that involve the sino-nasal cavities can be classified according to the presumed cause of the granuloma:

1. granulomas caused by infectious disease – actinomycosis, nocardia, tuberculosis, syphilis, rhinoscleroma, South American blastomycosis, leprosy, rhinosporidiosis, yaws, glanders, American mucocutaneous leishmaniasis;
2. granulomas caused by either autoimmune or collagen-vascular-related disease – Wegener's granulomatosis;
3. granulomas caused by lymphoma-related disease – midline granuloma;
4. granulomas caused by idiopathic process – sarcoidosis.

Wegener's granuloma is a nectrotizing granulomatous vasculitis that usually first affects the upper and lower respiratory tracts. Extensive local bone and soft-tissue destruction occurs, with erosion of the nasal septum and extension into the orbit. The localized form of the disease progresses to the systemic form with involvement of the kidneys and other organs.

Idiopathic midline granuloma is characterized by chronic necrotizing inflammation of the nose, sinuses, midline facial tissues and upper airways. This disease may be a lymphoma related illness.

Sarcoidosis affects the nasal cavities in 3% to 20% of patients with systemic sarcoidosis. The disease manifests with multiple small granulomas of the nasal septum. The paranasal sinuses are rarely affected.

Imaging characteristics of granulomas

Initially, there is nasal cavity involvement varying in severity from mucosal thickening to soft-tissue masses. There may be focal thickening of the nasal septum by a bulky soft-tissue mass (Figs 5.21 and 5.22). The maxillary and ethmoid sinuses are usually

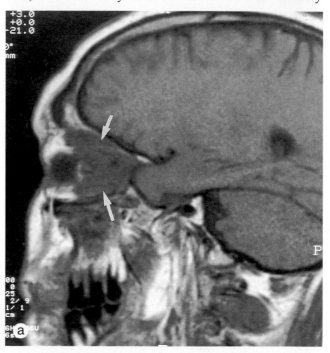

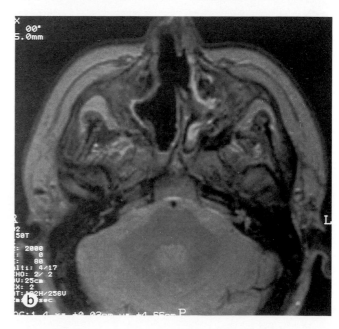

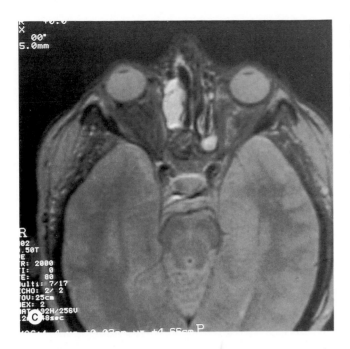

Fig. 5.21 Wegener's granuloma: orbits and sinuses.
MRI (**a**) sagittal T1W; (**b**) and (**c**) axial T2W. The dense granulomatous tissue in the orbits shows low to intermediate signal on both the T1W and T2W sequences (arrows) while the mucosa of the ethmoid sinuses shows high signal on T2W (**b**) and (**c**), characteristic of inflammation. (Reproduced by courtesy of Dr JPR Jenkins.)

affected after the nasal disease. The frontal sinuses are usually spared. There is a non-specific inflammatory mucosal thickening. The nasal bones and sinuses may be thickened, sclerotic sinus obliteration by reactive bone may occur [20] and there may be areas of bone erosion.

Plain radiographs may show bone destruction in the nose and sinuses and will demonstrate mucosal thickening or clouding of the antrum or other sinuses. CT is the definitive method of showing the degree of bone destruction present in both Wegener's and idiopathic midline granuloma. In midline granuloma the massive bone destruction may be demonstrated and this, with the absence of a soft-tissue mass, serves to distinguish it from neoplastic bone destruction, and is diagnostic. In Wegener's granuloma the bone destruction is less but it may be possible to show the irregular, ulcerated surface of the lining granulomatous tissue. As with other inflammatory conditions the granulomatous tissue gives high signal on T2-weighted sequences. Extensive fibrotic tissue may, however, be present, giving a low signal on both T1- and T2-weighted sequences.

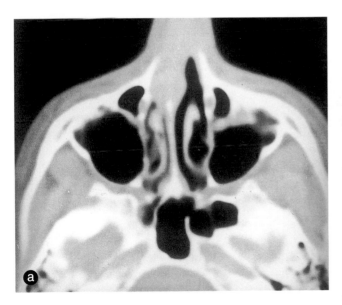

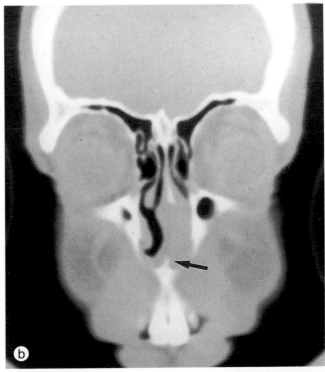

Fig. 5.22 Wegener's granuloma. CT scan (no contrast). (**a**) Axial plane; (**b**) coronal plane. A soft-tissue mass of granulation tissue is seen in the right nasal cavity with destruction of the nasal septum and early erosion of the anterior nasal spine (arrow).

CT AND FUNCTIONAL ENDOSCOPIC SURGERY [21]

Coronal scanning with the head in the prone position and hyperextended is the best position for pre-operative evaluation. CT is used to examine the osteo-meatal complex (the osteo-meatal complex refers to the maxillary sinus ostium, anterior and middle ethmoidal air cells, frontonasal duct, infundibulum and the middle meatal complex). The natural ostium of the maxillary sinus is important in endoscopic sinus surgery [22]. The ostium is located in the superior portion of the medial wall and drains into the posterior aspect of the ethmoid infundibulum as the sinus funnels into it, usually posterior to the middle point of the bulla ethmoidalis. The posterior extent of the uncinate process points to the position of the ostium and is an excellent imaging and endoscopic landmark for its localization. CT scanning and endoscopy are complementary in the diagnosis and treatment of disorders of the nasal cavity and paranasal sinuses (Fig. 5.23).

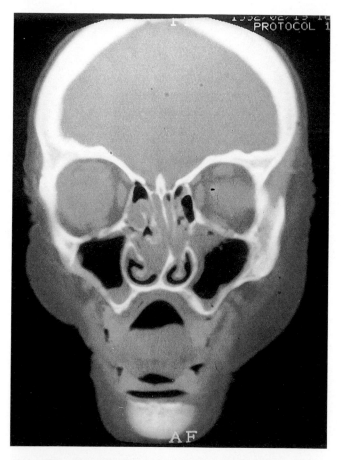

Fig. 5.23 Chronic sinusitis: pre-operative functional endoscopic surgery. Coronal CT scan (no contrast), showing moderate mucosal thickening bilaterally, worse on the left in the region of the osteomeatal complex.

CT technical factors are optimized with scanning in the prone position with thin 3 mm interval sections through the anterior paranasal sinuses. This allows optimal visualization of the osteo-meatal unit. The remaining posterior portions of the sinuses are adequately imaged using 5 mm interval slices. The coronal angle used is less critical. Exposure factors (mAs) can be reduced dramatically without image compromise. Data display is optimized when the bone algorithm is used to acquire the data and with the image display at intermediate window and width levels [21].

Role of CT in paranasal sinus inflammation

Coronal CT is the fastest and most effective means of identifying the affected sinuses. CT demonstrates small bony structures to great advantage and is the most useful modality for classifying acquired or congenital causes of paranasal sinus obstruction and localizing the region of the obstruction for endoscopic sinus surgery.

Role of MRI in paranasal sinus inflammation

MRI at present has a more limited role, but can be valuable:

1. for ruling out tumour in patients with isolated sinus opacification;
2. when findings suspicious of a primary tumour are identified on a CT scan, e.g. bone destruction;
3. for assessing intraorbital and intracranial complications.

MRI has specific signal characteristics for mucosal thickening and secretions, helps to demonstrate co-existing tumour and inflammation, and aids surgical planning and monitoring of response to treatment.

NEOPLASMS

Malignant tumours of the paranasal sinuses account for only 3% of tumours of the head and neck, 80% of these originating in the maxillary antra [23, 11]. Carcinomas of the paranasal sinuses carry a poor prognosis.

Role of imaging

1. Diagnosis
2. Differentiation of tumour from chronic inflammatory tissue
3. Tumour mapping
4. Staging
   TNM classification
   Demonstration of direct spread and nodal metastases
5. Post-treatment assessment
6. Detection of recurrence

Diagnosis

Conventional radiography, CT scanning and MRI are used to demonstrate the presence of bone erosion and an associated soft-tissue mass within the sinus cavities or outwith the confines of the sinuses. The presence of lymph node metastases may also be demonstrated on the sectional imaging. The imaging techniques must be optimal to demonstrate the subtle changes. In the case of CT thin sections are required to demonstrate subtle bone erosion which may be obscured by partial volume effect. Targeting of images and the selection of appropriate algorithms is essential. The use of high-dose intravenous contrast is necessary to demonstrate the tumour margins more accurately. Multiplanar MRI in the axial, coronal and sagittal planes is required and the use of paramagnetic contrast is mandatory for demonstrating tumour enhancement and for differentiating tumour from chronic inflammatory tissue.

The poor prognosis of malignant tumours of the paranasal sinuses is, in part, due to long delay between the development of symptoms and the establishment of the final diagnosis following biopsy. At diagnosis the tumour is frequently at an advanced stage, often inaccessible to radical surgical treatment. Furthermore, the tumour, in its early stages, may be obscured by accompanying chronic inflammatory disease.

In order to achieve earlier diagnosis the patient should be referred for imaging in the following two situations [24]:

1. when new symptoms such as epistaxis, facial pain and swelling, facial numbness and ocular dysfunction complicate known chronic sinusitis;
2. when conventional radiographic examination demonstrates an opacified sinus/soft-tissue mass and bone destruction.

The presence of a tumour of the paranasal sinuses is detected on imaging by demonstrating a soft-tissue mass with or without involvement of the surrounding bone enclosing the paranasal sinuses and nasal cavity. A small tumour contained within the sinuses without bone involvement presents as a non-specific mass on imaging.

Differentiation of tumour from chronic inflammatory tissue

Diagnostic difficulty can arise where chronic inflammatory tissue and tumour tissue have to be differentiated, as discussed previously. Chronic secretions may become desiccated, lose some of their signal intensity and exhibit intermediate signal intensity on both T1- and T2-weighted images. Thus, a chronically opacified sinus seen on a CT scan may demonstrate signal characteristics on MRI indistinguishable from a tumour. In this situation intravenous paramagnetic contrast is required for the differentiation. The enhancement pattern seen with the inflamed paranasal sinuses is predictable, exhibiting a peripheral rim of enhancement. A tumour shows an enhancing mass extending to the underlying bony margin.

Tumour mapping

The specific location of a maxillary carcinoma within the sinus can be correlated with the patient prognosis. The antrum is divided into a suprainfrastructure, mesoinfrastructure and an infrastructure with the lines of division being drawn on a coronal view of the sinuses through the antral floor and antral roof [25]. Using this system, tumours limited to the mesostructure and infrastructure require a partial or total maxillectomy, whereas tumours that involve the suprastructure require a total maxillectomy, and orbital exenteration. Furthermore, the antrum may be divided into posterosuperior and anteroinferior segments by a line drawn on a lateral view of the face from the medial canthus to the angle of the mandible (Ohngren's line). Tumours limited to the anteroinferior segment have a better prognosis.

Staging

Staging is based on the size and location of the tumour (T), the size and number of lymph node metastases (N) and the known presence of distant metastases (M) (TNM classification) [26].

The ultimate prognosis depends on the correct staging prior to surgical or irradiation therapy. Accurate staging depends on clinical evaluation, endoscopy with biopsy and imaging techniques. Multiplanar imaging (CT and MRI) improves staging in two ways.

1. Extension of tumours into clinically silent areas will be demonstrated.
2. Identification of the point of origin of the tumour can be assisted by defining the site of the greatest combined soft-tissue mass and osseous destruction.

There are no formal staging protocols for malignant tumours of the ethmoid, sphenoid and frontal sinuses. Malignant tumours of the maxillary sinuses can be staged according to the TNM classification (Table 5.1).

The lymphatic drainage of the maxillary sinus is via lateral and inferior collecting trunks to the submaxillary, parotid and jugulodigastric nodes and via a superoposterior trunk to the retropharyngeal and deep cervical nodes.

CT features of malignant tumours (Figs 5.24–5.26)

The less cellular tumours may appear non-homogeneous because of cystic degeneration, necrosis, or serous and mucous collection. The highly cellular tumour has a homogeneous appearance. Malignant tumours of the paranasal sinuses show irregular margins with evidence of thickened mucosa exhibiting variable enhancement characteristics depending on the volume and concentration of the contrast medium, type of tumour and cellularity and the timing of the post-contrast scan. Within a tumour, some areas demonstrate more enhancement, with tumour tissue easily differentiated from retained secretions or inflammatory tissue by the nature of the enhancement. Areas of tumour necrosis do not enhance.

The most characteristic sign of malignancy is the loss of surrounding fascial planes due to tumour infiltration. Extension of a soft-tissue mass outside the normal confines of a sinus cavity is an indication for biopsy.

CT is the method of choice for demonstrating destruction of the facial skeleton. Bone destruction should be displayed in axial and coronal planes. The destruction is usually lytic and there may be evidence of expansion of the sinus cavity with thinning of the margins or adjacent walls. Several bony fragments may be seen within the central portion of the mass due to the asymmetrical circumferential growth. Usually, the thin plates of bone are the first to be eroded. Sclerotic reaction is unusual and may be caused by concomitant chronic infection rather than tumour.

Table 5.1 Categories of malignant tumour

T categories

T1	Tumour confined to the antral mucosa of the infrastructure without bone erosion or destruction
T2	Tumour confined to the mucosa of the suprastructure without bone destruction or to the infrastructure with destruction of medial or inferior bony walls only
T3	Tumour invading the skin, orbits, anterior ethmoid sinuses, sphenoid sinuses, pterygoid plates, nasopharynx or skull base
T4	Tumour invading the cribriform plate, posterior ethmoid sinuses, sphenoid sinuses, pterygoid plates, nasopharynx or skull base

Nodal involvement

N0	No clinically positive node(s)
N1	Single clinically positive homolateral node 3 cm in diameter or less
N2	Single clinically positive homolateral node more than 3 cm but not more than 6 cm in diameter, or multiple homolateral nodes, none more than 6 cm in diameter
N3	Massive homolateral node(s), bilateral nodes or contralateral node(s)

Final staging

Stage I	T1 N0 M0
Stage II	T2 N0 M0
Stage III	T3 N0 M0
	T1, T2/T3 N1 M0
Stage IV	T4 N0/N1 M0
	Any T N2/N3 M0
	Any T any N M1

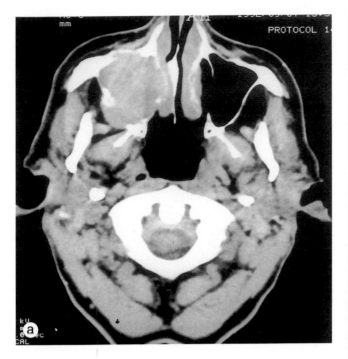

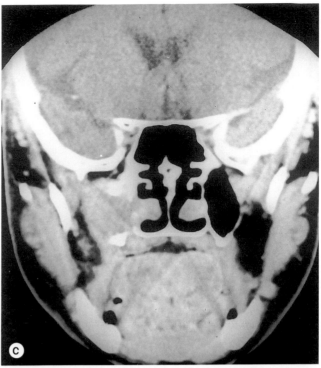

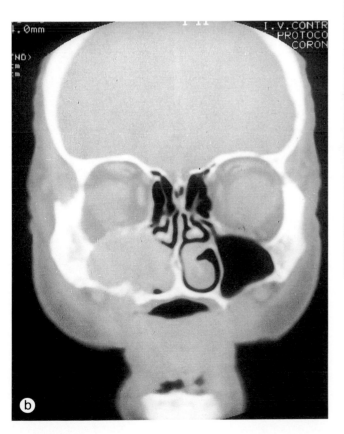

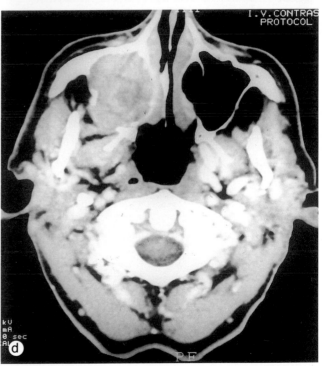

Fig. 5.24 Carcinoma of the maxillary antrum. CT scan – axial and coronal planes. Precontrast images (**a**), (**b**) and (**c**) and postcontrast image (**d**) detect a large mass of soft-tissue density in the right maxillary antrum with destruction of the anterior, lateral and medial walls. The mass extends into the nasal cavity. Note heterogeneous enhancement. There is slight extension into the infratemporal fossa and pterygopalatine region, with partial loss of the fat planes.

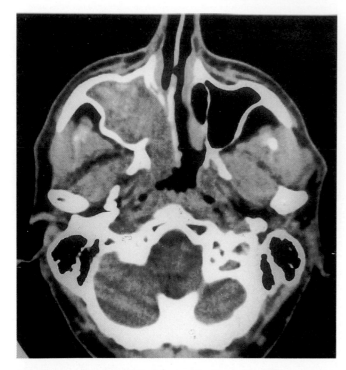

Fig. 5.25 Transitional cell carcinoma. CT scan (no contrast) – axial plane – through the maxillary antra shows a soft-tissue mass occupying the right antrum extending medially into the right nasal cavity and nasopharynx infiltrating the medial pterygoid muscle with loss of the tissue plane. Note the destruction of the medial wall of the antrum.

CT will demonstrate extension of the tumour into the pterygoid and infratemporal fossae, and into the parapharyngeal and pterygomaxillary regions. Contrast-enhanced CT scans are required for demonstrating:

1. early extension into the infratemporal and deep temporal fossae;
2. intracranial and orbital extension – demonstrated (Fig. 5.27) on direct coronal CT scans through the cribriform plate, frontal or sphenoid sinuses and the medial, lateral, superior and inferior walls of the orbit;
3. invasion of tumour into the adjacent vessels;
4. lymph node metastases in the submaxillary, retropharyngeal and jugulodigastric region.

Two varieties of orbital involvement may be shown.

1. More commonly, the orbital contents are displaced laterally without a direct invasion.
2. Less commonly, orbital invasion is a direct infiltration of the retro-orbital structures by tumour so that the normal anatomical landmarks are obliterated. This represents a later stage of orbital tumour growth.

Tumour may destroy the hard palate and extend into the contralateral nasal cavity.

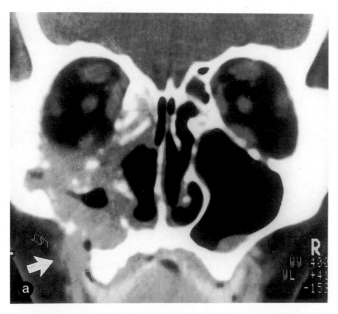

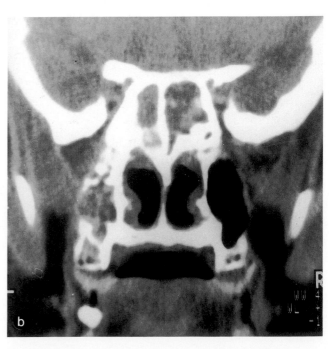

Fig. 5.26 Recurrent squamous cell carcinoma: left maxillary sinus. CT scan (no contrast). Extensive destruction of the margins of the antrum with intraorbital extension (**a**) and extension into the soft-tissue adjacent to the alveolar margin (arrow). Note the destruction of the pterygoid bone (**b**). The sphenoid sinus is opaque, with presumed retained secretions.

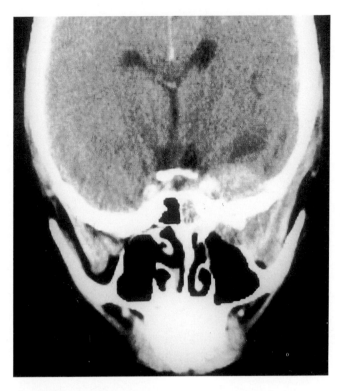

Fig. 5.27 Ethmoid carcinoma with intracranial extension. CT scan (postcontrast) – coronal plane – showing erosion of the skull base with enhancing tumour mass in the left middle cranial fossa. The left compartment of the sphenoid sinus is opaque due to either tumour tissue or retained secretions.

MRI features of malignant tumours (Figs 5.28 and 5.29)

Initially, it was anticipated that the tissue-specific signal characteristics of MRI would allow histological diagnosis. However, there is significant overlap of signal intensities between various tumours, precluding accurate distinction among malignancies or even between benign and malignant neoplasms within the paranasal sinuses.

Most tumours (80%) of the paranasal sinuses have intermediate signal on both T1 and T2-weighted images. This distinguishes them from most inflammatory processes, which have intermediate signal intensity on T1-weighted images but high signal intensity on T2-weighted images. Approximately 10%, including lymphomas and aesthesioneuroblastomas, have low to intermediate signal intensity on both T1- and T2-weighted images, reflecting their generally uniform, cellular histological pattern. The remaining 10% of malignant tumours, predominantly adenoid cystic carcinomas arising from the minor salivary glands, have a diverse histological pattern containing both serous and mucinous elements.

These display intermediate to high signals not only on T2- but also occasionally on T1-weighted images [13].

Because of the overlapping signal characteristics among tumour types and between tumours and inflammatory processes [27], the examination should include contrast administration where tumour is suspected. The presence of a homogeneously enhancing mass makes a tumour more likely. A recent review [13] has shown that enhanced MRI is more sensitive and more specific for differentiating neoplasms and inflammatory masses in the paranasal sinuses than is either non-contrast CT or unenhanced MRI.

Squamous cell carcinoma

This is the commonest malignant tumour of the paranasal sinuses (95%). In the nasal cavity the tumour usually arises in the middle turbinate and in some cases is linked to nickel exposure. Although 25–58% of the tumours arise in the maxillary antrum, the antrum is involved either directly or by extension in at least 80%. In 15–20% there is a history of chronic sinusitis and polyposis. The tumour is found more predominantly in males between the ages of 55 and 65 years.

These lesions are recognized on conventional imaging by the presence of variable soft-tissue masses in the sinus and by the osseous destruction that they cause in 70–90% of cases.

The treatment may be surgery, radiotherapy or both. Local recurrences are found in 20–50% and about 80% of these develop in the first year. Fifteen per cent develop nodal metastases at presentation and only 10% have distant metastases. The tumours are staged according to the TNM classification and the specific location may be divided into suprainfrastructure, mesoinfrastructure and infrastructure (see above).

Small carcinomas are often misdiagnosed as chronic sinusitis, nasal polyposis, lacrimal duct obstruction, tic douloureux or cranial arteritis. By the time of diagnosis, between 40% and 60% of patients have facial asymmetry, a tumour bulge in the oral cavity and tumour extension into the nasal cavity.

Primary frontal and sphenoid sinus carcinomas are rare. Frontal sinus carcinoma presents as acute frontal sinusitis. Cross-sectional imaging is performed to demonstrate the extent of the tumour. Carcinomas usually do not enhance on post-contrast CT scans. On MRI they have fairly homogeneous intermediate intensity on T1-, proton-density and T2-weighted images. Occasionally, localized areas of haemorrhage (high signal on T1- and T2-weighted images) or focal sites of necrosis (low signal on T1-, high signal on T2-weighted images) can be seen. Tumours show enhancement with paramagnetic contrast (Fig. 5.28).

The primary feature of these carcinomas is their strong tendency to destroy bone. Usually, the area of bone destruction is substantial compared with the size of the soft-tissue components. Bone remodelling is uncommon. The bone destruction is best shown on thin section CT scanning.

The location of the major arteries, veins and lymph nodes and determination of their involvement in the disease process are aided by contrast-enhanced dynamic scanning.

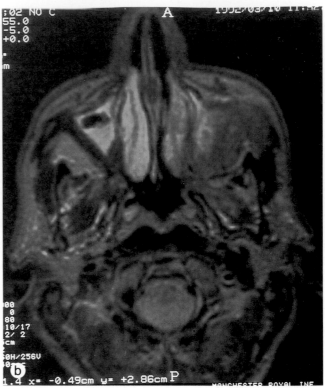

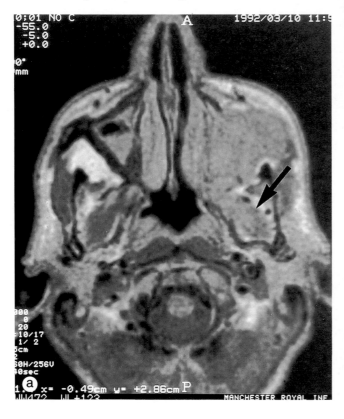

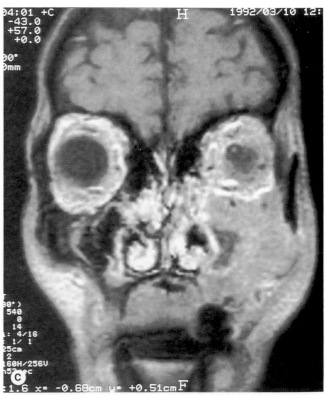

Fig. 5.28 Antral carcinoma. MRI. Axial proton-density (**a**) and T2W (**b**) spin echo images; coronal T1W gradient echo image (**c**) following gadolinium DTPA. Extensive tumour mass is seen in the left antrum but has clearly eroded outside the bony confines of the sinus. Note particularly the infiltration of the left lateral pterygoid muscle which is more easily seen on the proton-density image than the T2-weighted image (arrow in (**a**). In (**a**), the tumour is of similar intermediate signal to the inflammatory changes in the right antrum. On the T2W image (**b**), however, the inflammatory disease is of high signal while the tumour is of relatively low signal, permitting their differentiation. After contrast there is intense enhancement within the inflamed nasal mucosa but only minor tumour enhancement. (Reproduced by courtesy of Dr JPR Jenkins.)

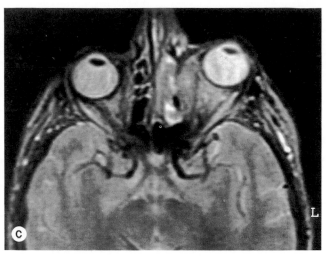

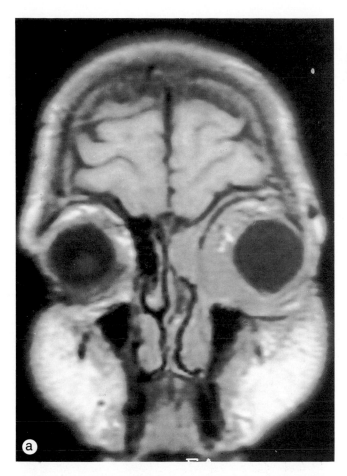

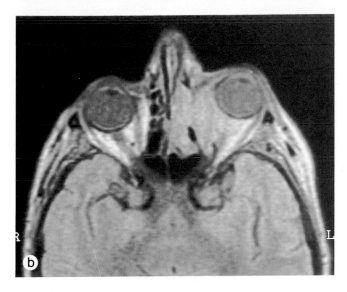

Fig. 5.29 Ethmoid carcinoma invading the orbit. MRI.
(**a**) Coronal T1W spin echo (no contrast); (**b**) axial proton-
density-weighted; (**c**) axial T2W spin echo. There is a mass
of intermediate signal intensity in the left ethmoid sinus,
extending into the infero-medial aspect of the left orbit and
displacing the globe anteriorly and laterally. Note the
obliteration of the retro orbital fat, shown best in (**a**). The
mass is confined to the ethmoid sinus and orbit.

Glandular tumours

These lesions account for 10% of all sinonasal
tumours and include adenoid cystic carcinoma,
mucoepidermoid carcinoma, acinous cell carcinoma,
benign and malignant pleomorphic adenoma and
adenocarcinoma. They vary from uniformly cellular
to highly pleomorphic tumours. They arise most
commonly in the palate and then extend into the
nasal fossae and sinuses.

Mucoepidermoid carcinomas and adenocarcinomas initially expand the sinus walls without erosion.
Eventually they will erode the bony walls. Similar
findings may be seen with skeletal metastases from
kidney, lung and breast.

Adenoid cystic carcinoma accounts for about 35% of
these tumours [28]. They arise from minor salivary
glands with a significant portion (26–41%) originating in the paranasal sinuses, nose or palate [29].
Forty-seven per cent arise in the maxillary sinuses,
32% involve the nasal fossae, 7% originate in the
ethmoid sinuses, 3% in the sphenoid sinuses, and
2% in the frontal sinuses [30]. These tumours
commonly occur in adults between the ages of 30
and 60 years and may recur 10–20 years after initial
treatment. Perineural spread with skip lesions is a
characteristic feature and complete surgical resection
cannot be guaranteed. The local post-surgical recurrence rate is 62% within 1 year, and 67–93% within 5
years. The worst prognosis is for antral tumours,
with 46% of patients being alive at 5 years but only
15% having no evidence of disease. Fifty per cent of
tumours have distant metastases to lungs, brain,
cervical, lymph nodes and bone. These tumours are
radiosensitive.

Adenocarcinoma may occur following exposure to hardwood and in the shoe industry. Most adenocarcinomas arise in the ethmoid sinuses and around the middle turbinates. They may be confused with metastases from colonic carcinoma.

Although adenocarcinomas have a lower tendency to metastasize to the cervical lymph nodes, they have a greater predilection for intracranial metastases.

Mucoepidermoid carcinoma is the third most common type of minor salivary gland tumour. They are of high-grade malignancy and resemble the adenocarcinoma. Most of these tumours involve the antrum and nasal cavity.

Benign mixed tumours are rare, mostly occurring in the nasal fossae. They usually arise from the nasal septum with the maxillary sinus the next most frequent site of origin. Wide surgical excision is required to prevent recurrence. Bone remodelling is typical. On MRI these tumours have an intermediate signal intensity on T1- and proton-density-weighted images. The T2 signal depends on the cellularity. Highly cellular lesions have an intermediate signal on T2-weighted images whereas less cellular lesions have a high signal.

Olfactory neuroblastoma or aesthesioneuroblastoma

These tumours arise from neurosensory receptor cells in the basal layer of the olfactory mucosa within the nasal cavities [31,32]. Continued growth leads to extension into the ethmoid and sphenoid sinuses or cranial cavity. The tumour is slow-growing and locally invasive, simulating both cranial and sinus neoplasia.

The peak incidences are 11–20 years and 50–60 years. On CT these tumours are homogeneous, enhancing masses that remodel bone (Fig. 5.30). They commonly extend into the ipsilateral ethmoid and maxillary sinuses and only rarely involve the sphenoid sinus. Calcification can occur [33]. On MRI they have intermediate signal characteristics on all imaging sequences.

Lymphoma

Lymphoma represents 8% of all paranasal sinus malignancies; 75% of these are non-Hodgkin's lymphoma. Between 2.2% and 6.5% of all lymphomas occur in the paranasal sinuses. Lymphomas are not

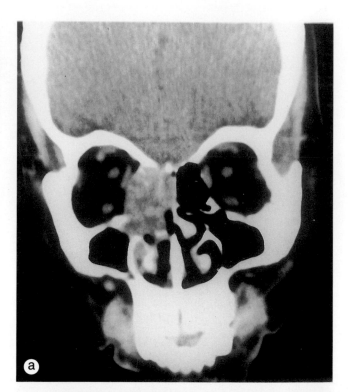

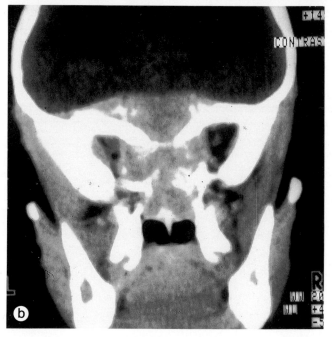

Fig. 5.30 Aethesioneuroblastoma. CT scan. Two cases: coronal sections. (**a**) Plain CT examination. Female aged 48. There is a soft-tissue mass in the right ethmoid sinus with destruction of the bony septae, medial orbital margin and nasal septum. The mass extends inferiorly into the nasal cavity. (**b**) Female aged 4 years. Postcontrast CT scan through the posterior ethmoid sinus shows a large enhancing mass with destruction of the ethmoid sinus, medial orbital margin and cribriform plate extending into the anterior cranial fossa. Small fragments of bone are seen within the enhancing mass (arrows).

associated with pre-existing nasal polyposis and sinusitis. These tumours are very radiosensitive and potentially curable. The most common cell type is histiocytic lymphoma. On imaging, they are bulky soft-tissue masses that enhance to a moderate degree. Bone remodelling tends to occur, with erosion of the bony margins (Fig. 5.31). They usually arise in the nasal fossae and maxillary sinuses, occurring less often in the ethmoid sinuses and rarely in the sphenoid and frontal sinuses. On MRI lymphomas have an intermediate-intensity signal on all sequences.

Burkitt's lymphoma is a special category. It occurs predominantly in childhood and in Central Africa and is related to the Epstein–Barr virus. Non-endemic Burkitt's lymphoma is rare and occurs in North America. It does not primarily involve the sinonasal cavities but spreads from the jaws, meninges, nasopharynx and lymph nodes.

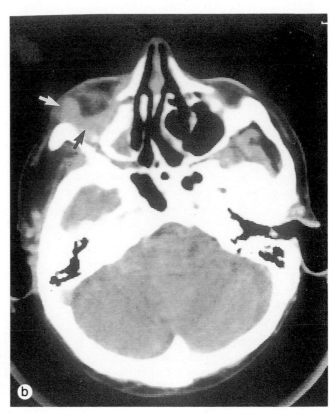

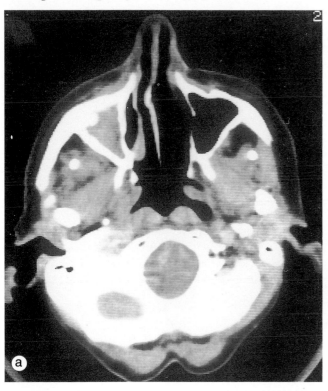

Fig. 5.31 Lymphoma of the right maxillary antrum and orbit. CT scan (postcontrast). CT sections at the lower (**a**) and upper (**b**) thirds of the maxillary sinus demonstrate complete sinus filling by lymphoma tissue. Spread into the lower anterior aspect of the right orbit has occurred where lobulated soft tissue is evident (arrows in **b**). On bone windows (**c**), erosion of the lower anterior sinus wall is present with spread of tumour into the adjacent subcutaneous fat (arrow). Five years previously the patient had presented with almost identical appearances but confined to the opposite side, and was successfully treated by radiotherapy.

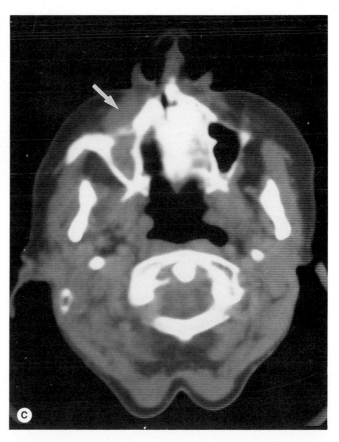

Rhabdomyosarcoma

This tumour accounts for 84% of all soft-tissue sarcomas and 35–45% of those occurring in the head and neck; 43% of those affected are less than 5 years of age. In the head and neck the most common sites in decreasing order for all tumours are the orbit, nasopharynx, middle ear and mastoids, sinonasal cavities, face, neck and larynx.

Although rhabdomyosarcomas can remodel bone or destroy it, on imaging most tumours show elements of both. On post-contrast CT these tumours show slight to moderate enhancement and are homogeneous in density. On MRI they are homogeneous and have intermediate signal intensities on all sequences.

Malignant melanoma

These are rare tumours, accounting for less than 2% of all melanomas, and comprise less than 3% of all sinonasal neoplasms. Malignant melanomas arising from the nasal septum are three times more common than those in the sinuses. In the paranasal sinuses the antrum is a site of origin in 80%. They rarely occur in the ethmoid sinuses. They tend to remodel bone but will also cause osseous destruction. Enhancement of the tumour mass is seen on post-contrast CT owing to the increased vascularity of the lesions. On MRI the tumour shows intermediate signal intensities on all sequences. However, in occasional cases high signal is seen on T1-weighted images due the presence of paramagnetic melanin. Malignant melanoma is treated by surgery and radiotherapy. Local recurrence occurs in 55–60% in the first year.

Other malignant tumours

Tumours of connective-tissue origin

These include malignant fibrous histiocytoma, fibrosarcoma, chondrosarcoma and osteosarcoma.

Fibrous histiocytoma

These tumours comprise an admixture of histiocytes and fibroblasts. They very rarely occur in the head and neck, mostly in the skin, orbit or sinonasal cavities. Local recurrence is common with cervical nodal metastases occurring in 12% and distant metastases in 42%. They show slight to moderate enhancement on post-contrast CT scans. Extensive

bone destruction occurs, resulting in appearances similar to a squamous cell carcinoma [34]. Similarly, on MRI these tumours have intermediate signal intensities on all imaging sequences.

Fibrosarcoma

Only 15% occur in the head and neck and most tumours involve the sinonasal cavities. They usually arise in patients between 20 and 60 years of age. Local recurrences develop 18 months after initial treatment and metastases occur within 2 years of local recurrences. These tumours have either a homogeneous or slightly non-homogeneous appearance on CT scanning and are non-enhancing following contrast administration. They remodel bone and on MRI have low to intermediate signal intensity on all imaging sequences.

Chondroma and chondrosarcoma

Chondrogenic tumours of the sinonasal cavities are rare and are mostly malignant. They are commonly found in middle-aged males. About 60% of the tumours arise in the anterior alveolar region of the maxilla. They present as a facial swelling or pain with loosening of the teeth or a nasal discharge or obstruction or epistaxis. Local recurrence is frequent following early, wide excision.

These tumours typically show stippled or amorphous parenchymal calcifications within an expanded sinus which shows remodelling of bone. However in many cases the calcifications may not be seen radiographically either on plain films or CT scans. In these cases the tumours have an attenuation value less than muscle, but greater than fat. They do not provoke sclerotic bone at their margins.

Benign tumours

Benign tumours of the paranasal sinuses are rare and include fibroma, angioma (Fig. 5.32), epithelial (inverted) papilloma, epidermoid (cholesteatoma), osteoma and osteochondroma, osteoclastoma and osteoid osteoma.

Osteoma

The osteoma is a benign proliferation of bone and is the commonest benign tumour affecting the paranasal sinuses. They show a predominantly male sex distribution. Osteomas are of two types: the more common ivory osteoma, which consists of hard dense bone, and the cancellous osteoma. These

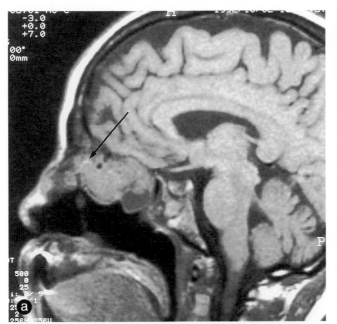

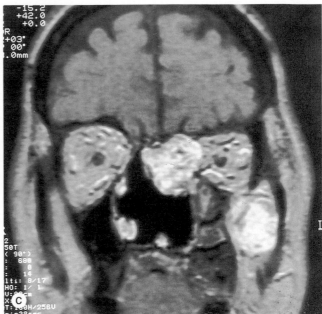

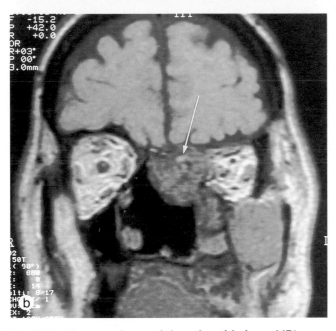

Fig. 5.32   Haemangioma of the ethmoid sinus. MRI sagittal spin echo sequence (**a**); coronal T1W graded echo sequence precontrast (**b**) and following gadolinium DTPA (**c**). The patient had undergone partial tumour resections several years previously. Heterogeneous tumour is seen in the left ethmoidal region but is expanding across the midline as well as into the left orbit. Tumour has extended into the left antrum and further laterally into the infratemporal region, and enhances strongly after contrast (**c**). Three smaller enhancing nodules are seen in the expanded nasal cavity (**c**). Note the multiple areas of flow-related signal void within the ethmoidal mass on both sequences. Focal areas of high signal on the T1W sequence (**a**) adjacent to the flow voids from previous haemorrhage (arrows) are present. (Reproduced by courtesy of Dr A Jackson.)

tumours occur mainly in the frontal sinuses followed in descending order by the ethmoid, maxillary and sphenoid sinuses. The high prevalence of osteomas in the frontal and ethmoid sinuses may relate to the fact that this region is the junction of membranous and enchondral development of the frontal and ethmoid bones.

Bone density varies from a very dense, sclerotic lesion in the ivory osteoma to a progressively less dense and less ossified lesion in the cancellous type. They have a round or lobulated outline. The affected sinus is otherwise normal in appearance, unless the osteoma obstructs the ostium resulting in a secondary mucocele. Very large osteomas can cause erosion of the walls of the sinus with encroachment on the orbit and cranium. Osteomas are the most common benign paranasal sinus tumour to be associated with spontaneous cerebrospinal fluid rhinorrhoea [35]. In Gardener's syndrome multiple osteomas are seen.

Osteomas are often incidental findings on plain radiographs.

On CT scanning osteomas arise from one of the sinus walls (Fig. 5.33) or the intersinus septum. The differences between the types of osteoma correlate with the degree of bone density within the tumour. On MRI these lesions give a non-homogeneous, low to intermediate signal density on all imaging sequences. Based on MRI findings, the osseous nature of the tumour may go undetected.

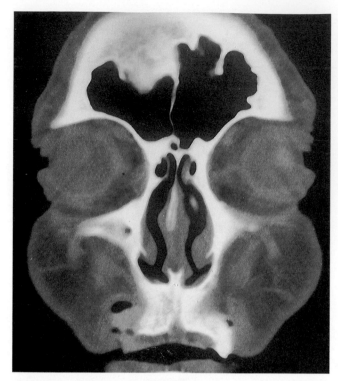

Fig. 5.33 Osteoma: frontal sinus. Coronal CT scan (no contrast) shows well-defined mass of bone density arising from the roof of the right compartment and projecting into the air-filled sinus.

Epithelial (inverted) papilloma

This is a rare tumour arising from the mucous membrane of the nasal cavity, less commonly from the paranasal sinuses. It occurs predominantly in males. There is a tendency to recurrence after removal and malignant change occasionally takes place. The imaging features usually consist of a mass in the nasal cavity with opacification of the paranasal sinuses. The mass produces expansion of the nasal cavity and thinning of its walls. Where the ostium is obstructed the affected sinus is opaque. However, if the tumour arises in the maxillary antrum the tumour mass completely occupies the sinus with expansion and thinning of the walls. The appearances may be confused with a mucocele or even a carcinoma. Biopsy is necessary to exclude malignancy [36].

Fibrous dysplasia

Bones of the head and neck comprise 20–25% of the cases of monostotic fibrous dysplasia, with the maxilla and mandible being the most common sites. Conventional radiography and CT scanning show thickening of the facial skeleton with encroachment on the lumen of the maxillary sinus. Albright's syndrome consists of polyostotic fibrous dysplasia, cutaneous pigmentation and sexual precocity.

Leontiasis ossea is the descriptive term when extensive involvement of the facial bones with resulting distortion of the face, including invasion of the orbit, occurs.

Cholesteatoma (epidermoid)

This lesion occurs occasionally in the maxillary antra and may fill the sinus cavity with expansion of its walls. The radiographic features are similar to those of a mucocele. CT or MR scanning may be of value in identifying the fat content of a cholesteatoma.

Odontogenic tumours

Common odontogenic tumours involving the paranasal sinuses are ameloblastoma, cementoma, odontoma and fibromyxoma.

Ameloblastoma

This tumour is derived from the odontogenic apparatus but does not form a hard tumour. Most tumours occur in the third and fourth decades of life and 90% of the maxillary lesions involve the premolar–molar area. The tumour presents as a painless swelling with nasal obstruction if the maxilla is involved. On plain radiographs the tumour appears as a multioculated, lytic lesion. If the tumour extends into the antrum, the sinus cavity is clouded and the sinus walls are remodelled and destroyed. On CT scanning these tumours tend to have a non-enhancing, non-homogeneous appearance and on MRI they have non-homogeneous mixed signal intensity. On T1- and proton-density-weighted images these tumours display intermediate intensity signals while on T2-weighted sequences the tumours have variable intermediate and high signal intensity.

Complete surgical excision is the treatment of choice.

Fibromyxoma

These tumours tend to occur in the second and third decades of life. In the maxilla and paranasal sinuses they can be locally aggressive and have a high recurrence rate. On imaging these tumours are primarily expansile when they involve the sinuses although focal areas of bone destruction can occur. These lesions usually have flecks of calcification within the tumour matrix.

Metastases to the paranasal sinuses

Renal cell carcinoma is the most common primary tumour metastasizing to the paranasal sinuses and in some cases the metastases may precede the diagnosis of the primary tumour. Next in frequency are tumours of the lung and breast followed by those of the testis, prostate and gastrointestinal tract. Squamous and basal cell carcinoma of the face and scalp may metastasize to the sinuses with perineural tumour invasion.

On imaging, metastases from primary renal cell carcinomas enhance and may cause remodelling of the bony margins of the sinuses as well as producing bony destruction. Metastases from the other sites are aggressive tumours with large non-enhancing soft-tissue components and extensive bone destruction.

POST-TREATMENT ASSESSMENT

Post-operative imaging of the paranasal sinuses may be difficult to interpret. The detection of tumour recurrence and the degree of perineural and perivascular extension must be assessed for accurate staging. Some tumours, e.g. adenoid cystic carcinoma, extend along the nerves and may produce skip lesions with intervening normal tissue between the primary site and the site of recurrence. While CT can demonstrate bone destruction, particularly on comparison with a pre-operative examination, assessment of soft-tissue tumour recurrence by means of CT scanning is unreliable. In the absence of bone destruction the diagnosis of residual or recurrent tumour should be made before bone destruction occurs. CT cannot differentiate recurrent tumour from scar tissue or associated inflammatory changes.

Contrast-enhanced MRI may show extension of the soft-tissue mass more accurately than CT but will still be non-specific in differentiating recurrent tumour from active scar or irradiated tissue.

The post-operative appearance of the antral cavity following the resection of a malignancy depends on the time interval between surgery and the sectional imaging examination. During the first 2–4 weeks after an operation the surgical cavity may have irregular margins caused by healing and/or inflammation but 6–8 weeks post-operatively the margins should be smooth. Any nodular irregularity in the margins indicates fibrous mucosal scarring or recurrent disease and is an indication for biopsy [11]. More advanced recurrent disease appears as an expanding mass on post-operative sectional imaging, particularly contrast-enhanced MRI (Fig. 5.34).

Finally, CT and MRI can be used to monitor the results of irradiation and chemotherapy [37]. Bone destruction may return to normal or there may be a decrease in the size of the tumour masses.

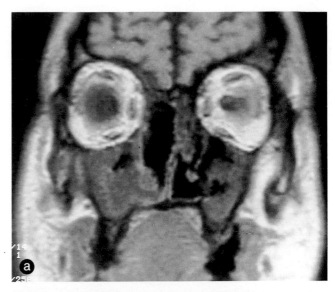

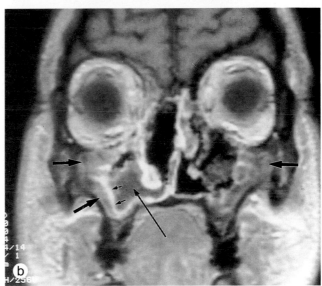

Fig. 5.34 Recurrent papillary adenocarcinoma: maxillary antra and ethmoid sinuses. MRI. Coronal T1W gradient echo images (**a**) before and (**b**) following gadolinium DTPA. Ethmoidal papillary adenocarcinoma was originally diagnosed 6 years earlier and treated with surgery and radiotherapy. The unenhanced scan (**a**) shows extensive, slightly heterogeneous low signal changes in both maxillary antra, the nasal cavity and left ethmoid sinuses. After contrast (**b**) three components of differing signal are evident. Heterogeneously enhancing tumour of generally intermediate signal is seen along the periphery of both antra (large, short arrows), displacing intensely enhancing inflamed mucosa (small arrows) away from the signal void of the antral walls. In the centre of the right antrum is an area of almost homogeneous, non-enhancement consistent with inflammatory fluid or debris (long arrow). Similar changes are seen in the left ethmoid region. Biopsy confirmed recurrent tumour. Multiple areas of bone deficiency are present, due to both surgery and tumour erosion.

REFERENCES

1. Bilaniuk LI, Zimmerman RA. Computed tomography in evaluation of the paranasal sinus. Radiol Clin N Amer 1982; 20: 51–66.
2. Lloyd GA. CT of the paranasal sinuses: study of a control series in relation to endoscopic sinus surgery. J Laryngol Otol 1990; 104(6): 477–481.
3. Duvoisin B, Agrifoglio A. Prevalence of ethmoid sinus abnormalities on brain CT of asymptomatic adults. AJNR 1989; 10(3): 599–601.
4. Som PM, Curtin HD. Chronic inflammatory sinonasal diseases including fungal infections. The role of imaging. Radiol Clin N Amer 1993; 31: 33–44.
5. Som PM. The paranasal sinuses. In: Bergeron RT, Osborn AG, Som PM, eds. Head and neck imaging excluding the brain. St Louis: CV Mosby, 1984.
6. Som PM, Shugar JMA. The CT classification of ethmoid mucoceles. J Comput Assist Tomography 1980; 4: 199.
7. Jacobs M, Som PM. The ethmoidal 'polypoid mucocele'. J Comput Assist Tomography 1982; 6: 721.
8. Naul LG, Hise JH, Ruff T. CT of inspissated mucus in chronic sinusitis. AJNR 1987; 8: 574.
9. Som PM, Sacher M, Lawson W, Biller H. CT appearance distinguishing benign nasal polyps from malignancies. J. Comput Assist Tomography 1987; 11; 129–133.
10. Som PM et al. Ethmoid sinus disease: CT evaluation in 400 cases. Part 1: Non-surgical patients. Radiology 1986; 159: 591.
11. Som PM, Shugar JMA, Biller HF. The early detection of antral malignancy in the post-maxillectomy patient. Radiology 1982; 143: 509–512.
12. Van Tassel P, Lee YY, Jing BS et al. Mucoceles of the paranasal sinuses: MR imaging with CT correlation. AJNR 1989; 10: 607–612.
13. Lanzieri CF, Shah M, Krauss D, Lavertu P. Use of gadolinium-enhanced MR imaging for differentiating mucoceles from neoplasms in the paranasal sinuses. Radiology 1991; 174: 425–428.
14. Kaufman DM, Litman N, Miller MH. Sinusitis-induced subdural empyema. Neurology 1983; 33: 123.
15. Williams HL. Infections and granulomas of the nasal airways and paranasal sinus. In: Paparella MM, Shumrick DA, eds. Otolaryngology, vol 3: Head and neck. Philadelphia: WB Saunders, 1973: p 27–32.
16. Centeno RS, Bentson JR, Mancuso AA. CT scanning in rhinocerebral mucormycosis and aspergillosis. Radiology 1981; 140: 383.
17. Robb PJ. Aspergillosis of the paranasal sinuses: a case report and historical perspective. J Laryngol Otol 1986; 100: 1071–1077.
18. Patel PJ, Kolawole TM, Malabarey A et al. CT findings in paranasal aspergillosis. Clin Radiol 1992; 45: 319–321.
19. Demaerel P, Brown P, Kendall BE, Revesz T, Plant G. Case report: allergic aspergillosis of the sphenoid sinus: pitfall on MRI. Br J Radiol 1993; 66: 260–263.
20. Paling MR, Roberts RL, Fauci AS. Paranasal sinus obliteration in Wegener's granulomatosis. Radiology 1982; 144: 539.
21. Babbel R, Harnsberger HR, Nelson B et al. Optimization of techniques in screening CT of the sinuses. AJNR 1991; 12: 849–854.
22. Mafee MF. Preoperative imaging anatomy of the nasal-ethmoid complex for endoscopic sinus surgery. Radiol Clin N Amer 1993; 31: 1–20.
23. Kondo M, Masatoshi H, Shiga H et al. Computed tomography of malignant tumours of the nasal cavity and paranasal sinuses. Cancer 1982; 50: 226–231.
24. Hasso AN. CT of tumours and tumour-like conditions of the paranasal sinuses. Radiol Clin N Amer 1984; 22(1): 119–130.
25. Baredes S, Cho HT, Som ML. Total maxillectomy. In: Blitzer A, Lawson W, Friedman WH, eds. Surgery of the paranasal sinuses. Philadelphia: WB Saunders, 1985: p 204–216.
26. Baker HW. Staging of cancer of the head and neck: oral cavity, pharynx, larynx and paranasal sinuses. Cancer 1983; 33: 130–138.
27. Som PM, Dillon WP, Sze G et al. Benign and malignant sinonasal lesions with intracranial extension: differentiation with MRI. Radiology 1989; 172: 763–766.
28. Spiro RH et al. Tumours of minor salivary origin; a clinico-pathologic study of 492 cases. Cancer 1973; 31: 117.
29. Barnes L, Verbin RS, Gnepp DR. Disease of the nose, paranasal sinuses, and nasopharynx. In: Barnes L, ed. Surgical pathology of the head and neck, vol 1. New York: Marcel Dekker, 1985: p 403–451.
30. Miller RH, Calcaterra TC. Adenoid cystic carcinoma of the nose, paranasal sinuses and palate. Arch Otolaryngol 1980; 106: 424–426.
31. Burke DP, Gabrielson TO, Knake JE et al. Radiology of olfactory neuroblastoma. Radiology 1980; 137: 367–372.
32. Hurst RW, Erickson S, Cail WS et al. Computed tomographic features of esthesioneuroblastoma. Neuroradiology 1989; 31(3): 253–257.
33. Regenbogen VS et al. Hyperostotic esthesioneuroblastoma: CT and MR findings. J Comput Assist Tomography 1988; 12: 52.
34. Merrick RE, Rhone DP, Chilis TJ. Malignant fibrous histiocytoma of the maxillary sinus. Arch Otolaryngol 1980; 106: 365.
35. Shugar JMA et al. Non-traumatic cerebrospinal fluid rhinorrhea. Laryngoscope 1981; 41: 114.
36. Lund VJ, Lloyd GAS. Radiological changes associated with inverted papilloma of the nose and paranasal sinuses. Br J Radiol 1984; 57: 455–461.
37. Raney, RB Jr, Zimmerman RA, Bilaniuk LT et al. Management of cranio-facial sarcoma in childhood assisted by computed tomography. Radiat Oncol Biol Phys 1979; 5: 529–534.

6 *The nasopharynx, oropharynx and oral cavity*

Bernadette M. Carrington and Richard J. Johnson

ANATOMY

Understanding naso- and oropharyngeal anatomy [1–9] is important for accurate interpretation of cross-sectional images, particularly in the field of oncology. In-depth anatomical description of the entire region is beyond the scope of this chapter and the reader is referred to referenced works for more detailed information. Discussion will concentrate on the anatomical features and compartments which are important in the evaluation of pharyngeal pathology.

General pharyngeal anatomy

The pharynx [1,2] is a tubular structure extending from the base of skull down to the level of C6 vertebral body and the inferior border of the cricoid cartilage. It is subdivided into three compartments, the nasopharynx, oropharynx and hypopharynx (Fig. 6.1).

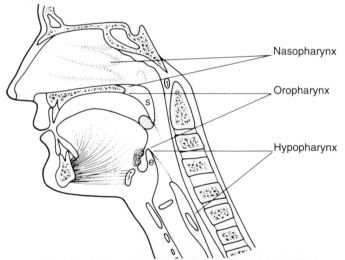

The pharyngeal wall is composed of four layers: a mucous membrane, fibrous layer, pharyngeal muscular layer and a thin outer fascial covering, the buccopharyngeal fascia.

The mucous membrane is identified on MR images (Fig. 6.2) but often cannot be separated from the rest of the pharyngeal wall on CT studies. The fibrous layer, the pharyngobasilar fascia, is thickened superiorly in the upper nasopharynx and can be identified on MR scans (Fig. 6.2). It diminishes in thickness inferiorly but part of it remains visible posteriorly as the median pharyngeal raphe, into which the intrinsic muscles of the pharynx insert. There are three intrinsic pharyngeal muscles, the superior, middle and inferior constrictors, plus three extrinsic muscle pairs, the stylopharyngei, salpingopharyngei and palatopharyngei which are more vertically orientated and pass into the pharyngeal wall.

The buccopharyngeal fascia is not normally visible on cross-sectional imaging studies.

Arterial supply

The pharynx is supplied by the ascending pharyngeal branch of the external carotid artery, plus branches of the facial, maxillary and lingual arteries.

Venous drainage

A pharyngeal venous plexus communicates superiorly with the pterygoid venous plexus and drains inferiorly into the internal jugular and facial veins.

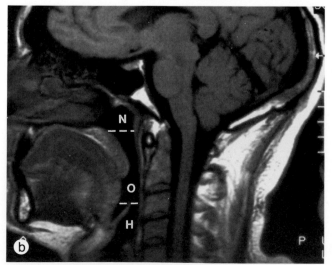

Fig. 6.1 Divisions of the pharynx. (a) Line diagram. **(b)** Sagittal T1-weighted MR image. In **(a)**, dotted areas indicate the position of the pharyngeal, palatine and lingual tonsils.
N = nasopharynx; O = oropharynx; H = hypopharynx; S = soft palate; e = epiglottis.

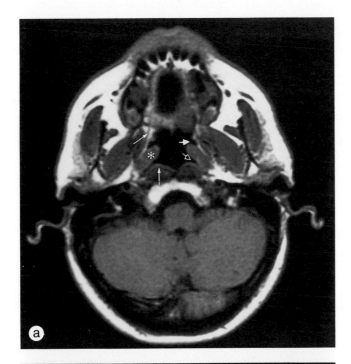

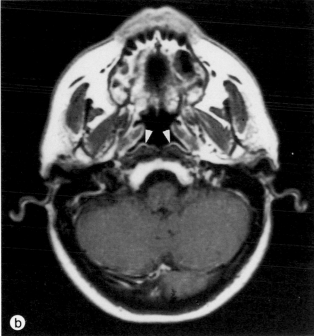

Fig. 6.2 Normal MR anatomy of the pharyngeal wall, before and following administration of gadolinium DTPA. (**a**) T1-weighted precontrast transaxial image. (**b**) T1-weighted postcontrast transaxial image. The mucous membrane (arrowheads in (**b**)) is of high signal intensity on T1-weighted images and enhances after gadolinium. The pharyngobasilar fascia (long white arrows) is of low signal intensity pre- and postcontrast and is situated immediately beneath the mucosa posteriorly. Laterally it encompasses the levator veli palatini (asterisk) within the pharyngeal musculature. The eustachian tube orifices (small white arrow) and lateral pharyngeal recesses (small open arrow) are well seen.

Lymphatics

Some areas of the pharynx initially drain through outlying lymph nodes groups but *all* efferent pharyngeal lymphatics ultimately pass to the deep cervical lymph nodes. The deep cervical lymph nodes are situated along the length of the carotid sheath, which surrounds the internal carotid artery and internal jugular vein. They are divided into superior and inferior nodal groups by the omohyoid muscle (Fig. 6.3).

The superior deep cervical lymph nodes are located deep to the sternocleidomastoid muscle adjacent to the upper internal jugular vein and lateral to the carotid artery. They are impalpable clinically except for one identifiable subgroup, the jugulo-digastric nodes, which are located between the facial and internal jugular veins and the posterior belly of the digastric muscle.

The inferior deep cervical lymph nodes are situated along the lower internal jugular vein and, more inferiorly, are related to the brachial plexus and subclavian vessels. The jugulo-omohyoid node, situated on the omohyoid tendon, is a member of the inferior deep cervical nodal group and is one of the principal drainage nodes for the tongue.

Efferents from each deep cervical lymph node chain form a jugular trunk. The left jugular trunk drains into the thoracic duct and the right jugular trunk drains into the junction of the right internal jugular and subclavian veins.

The nasopharynx

The nasopharynx [1–4] is that part of the pharynx situated above the soft palate. Its borders comprise the clivus superiorly, and the soft palate and opening of the oropharynx inferiorly. Anteriorly the nasopharynx communicates with the nasal cavity and posteriorly it is separated from the clivus and anterior arch of the atlas by a potential space, the retropharyngeal space. Laterally, the nasopharynx is related to the parapharyngeal spaces. Several important structures are found in the nasopharynx, namely the pharyngeal opening of the eustachian tubes, the tubule elevations, the pharyngeal recesses and the pharyngeal tonsil.

Each eustachian tube opens into the lateral wall of the nasopharynx at the same horizontal level as the inferior nasal concha (Fig. 6.2). The eustachian tube orifices appear symmetrical and are airfilled for the first 3–4 mm. The orifices are bounded superiorly and posteriorly by the tubule elevations (tori tubarii) representing the cartilaginous continuation of each

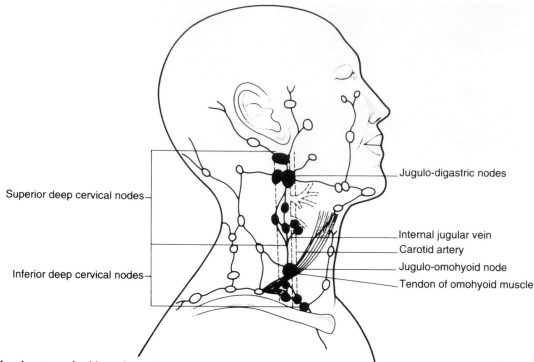

Fig. 6.3 The deep cervical lymph nodes.

eustachian tube. These are easily seen on cross-sectional imaging where they appear slightly asymmetrical in size and contour. Two folds of mucosa extend inferiorly from the tubule elevations. One runs inferiorly and posteriorly into the pharynx and is called the salpingopharyngeal fold since it overlies the salpingopharyngeus muscle. The other runs antero-inferiorly and covers part of the tensor veli palatini muscle. It is called the nasopalatine fold.

The pharyngeal tonsil is largest in children up to 6–7 years of age and then atrophies. It lies immediately inferior to the clivus. The lymphoid tissue within the tonsil is called the adenoids.

Lateral pharyngeal recesses (fossae of Rosenmuller) are situated above and posterior to the eustachian tubule elevations and salpingopharyngeal folds (Fig. 6.2). Each lateral pharyngeal recess lies directly inferior to the foramen lacerum and carotid canal. Intracranial tumour spread can occur into the cavernous sinuses and middle cranial fossa from the lateral pharyngeal recesses, which are the commonest site of nasopharyngeal tumours. The lateral pharyngeal recesses may not be identified in children, since the pharyngeal tonsil may extend into and obliterate the recesses.

Lymphatic drainage

The nasopharynx drains initially to regional outlying retropharyngeal nodes situated between the bucco-pharyngeal and prevertebral fascia. These comprise a central median nodal group, plus two lateral groups which are situated in front of the lateral masses of the upper cervical vertebrae and lateral to the longus capitis muscles (Fig. 6.4). Efferent retropharyngeal vessels drain into the deep cervical nodes. The nasopharynx also has direct lymphatic drainage into the deep cervical nodes and a third drainage pathway occurs via the spinal accessory nodes, which lie along the line of the spinal accessory nerve.

The oropharynx

The oropharynx [1,2,5,6] extends from the soft palate down to the upper border of the epiglottis. Its boundaries are the soft palate and nasopharyngeal opening superiorly and the anterior surface of the epiglottis and vallecular mucosa inferiorly. Anteriorly the oropharynx communicates with the oral cavity via the oropharyngeal opening; its posterior relations are C2 vertebral body plus the upper half of C3 vertebral body. The lateral walls of the oropharynx directly adjoin the parapharyngeal spaces and the carotid sheaths.

Major anatomical structures within the oropharynx are the palatine tonsils and tonsillar pillars together with the base of tongue, which includes the lingual tonsil.

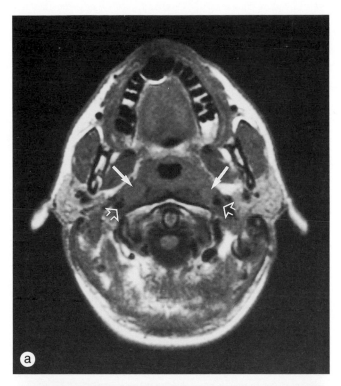

The palatine tonsils are aggregates of lymphatic tissue surrounded by a fibrous capsule, which are located on the lateral oropharyngeal wall between the anterior and posterior tonsillar pillars (Fig. 6.5). The anterior tonsillar pillars or palatoglossal folds are formed by the palatoglossal muscles and overlying mucous membrane. The posterior tonsillar pillars are composed of the palatopharyngeal muscles with their overlying mucosa. The anterior and posterior tonsillar pillars diverge and form the glossotonsillar (glossopharyngeal) sulci in the low oropharynx.

The base of the tongue is the posterior third of the tongue, which commences at the level of the circumvallate papillae and extends postero-inferiorly to the valleculae. It is formed by the intrinsic tongue muscles and two extrinsic muscles, the styloglossus and hyoglossus. The hyoglossus muscle is an important anatomical structure on coronal CT and MRI scans, since the lingual artery passes through the base of the tongue medial to the muscle, while the deep lingual vein and hypoglossal nerve pass lateral to it.

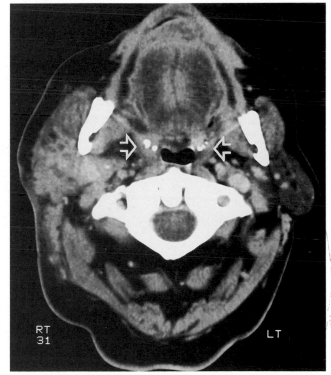

Fig. 6.4   MRI demonstrating enlarged lateral retropharyngeal nodes in a patient with nasopharyngeal carcinoma. (a) T1-weighted precontrast transaxial image. (b) T1-weighted postcontrast transaxial image. Bilateral retropharyngeal nodes are seen (white arrows) situated anterior to the lateral masses of the vertebral body and lateral to the longus capitis muscles (asterisks in (b)). The nodes lie medial to the carotid sheaths (open arrows).

Fig. 6.5   The palatine tonsils. Transaxial CT scan showing the palatine tonsils (arrows) containing tonsilliths. The right tonsil is mildly enlarged. There is metastatic lymphadenopathy involving the right parotid gland.

The lingual tonsil is composed of lymphoid tissue forming raised nodules or ridges beneath the base of tongue mucosa. It is of variable size and appearance and is asymmetrical on imaging in most patients (Fig. 6.6).

A posterior midline fold, the glossoepiglottic fold, separates the two valleculae, and the paired lateral pharyngoepiglottic folds separate the tongue and valleculae anteriorly from the pyriform sinuses and hypopharynx posteriorly.

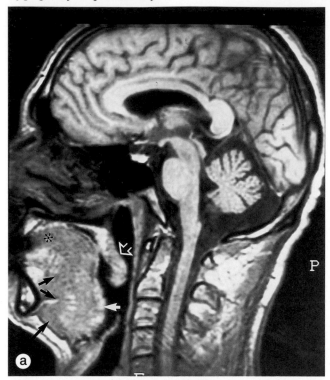

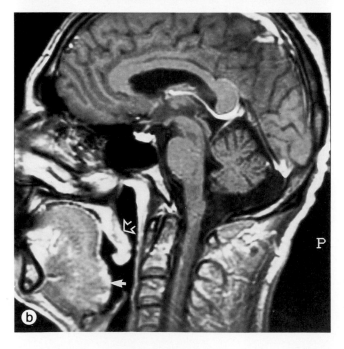

The oral cavity

The oral cavity [1,2,6] is situated anterior to the oropharynx and separated from it by the oropharyngeal opening, which comprises the soft palate, anterior tonsillar pillars and circumvallate papillae. The superior boundary of the oral cavity is the hard palate, together with the superior alveolar ridge and teeth, while the inferior boundary is formed by the floor of the mouth plus the inferior alveolar ridge and teeth. The lateral margin of the oral cavity is demarcated by the cheek, which consists of the buccinator muscle and its overlying mucosa. Important contents of the oral cavity are the anterior two-thirds of the tongue, the floor of the mouth, the hard palate, the upper and lower gingivobuccal sulci and the retromolar trigone.

The anterior two-thirds of the tongue comprises an outer mucosal covering and both intrinsic and extrinsic muscles (Fig. 6.7). The intrinsic muscles of the tongue are divided into superior and inferior longitudinal, plus transverse and vertical muscle bundles. They are not easily separated on imaging studies but the longitudinal fibres may be identified on MR (Fig. 6.6). The extrinsic muscles are the genioglossus, geniohyoid and hyoglossus muscle pairs, which are well seen on coronal imaging (Fig. 6.7). Both the genioglossus and geniohyoid muscle pairs are separated by a fatty midline lingual septum, which also partially divides the inferior portion of the intrinsic tongue muscles. This septum is symmetrical and easily seen on CT or MRI (Fig. 6.7). Its obliteration gives important information about tumour spread through the tongue.

The floor of the mouth consists of the mylohyoid muscles which slope downward and medially from their origin on the mylohyoid line of the medial mandible to form a midline raphe (Fig. 6.7). Posteriorly, some slips of the mylohyoid muscle insert into the hyoid bone. The mylohyoid muscles delineate the margin of two important anatomical spaces in the floor of the mouth, the sublingual and submandibular spaces.

Fig. 6.6 Normal sagittal anatomy of the oropharynx and oral cavity. Sagittal T1-weighted MR images (**a**) precontrast and (**b**) postcontrast. The genioglossus muscle (small black arrows) is seen forming the bulk of the tongue with the longitudinal intrinsic muscles (asterisk) situated superiorly and the geniohyoid muscle (long black arrow) inferiorly. The lingual tonsil (white arrow) is of high signal intensity and causes an irregular, corrugated appearance to the tongue base. After contrast the lingual tonsil and soft palate (open arrow in (**b**)) enhance markedly.

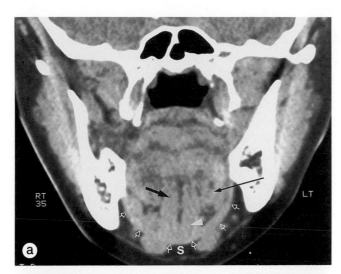

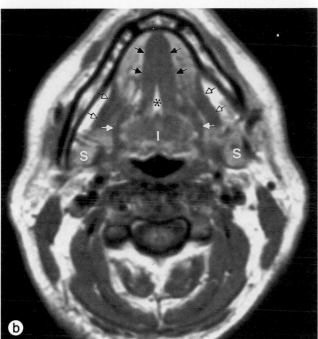

Fig. 6.7   The anterior two-thirds of the tongue and floor of the mouth. (a) Coronal CT image demonstrating the hyoglossus/styloglossus muscle complex (long black arrow), the genioglossus muscles (short black arrow), the geniohyoid muscles (white arrowhead) and the mylohyoid sling (small open arrows). The submental space (S) is seen beneath the mylohyoid sling between the anterior bellies of the digastric muscles. **(b)** Transaxial T1-weighted MR image demonstrating the fatty midline lingual septum (asterisk), delineated by paramedian genioglossus muscles (small black arrows). Posteriorly, the styloglossus/hyoglossus muscle complex (small white arrows) demarcates the lateral margin of the intrinsic muscles of the tongue.

The sublingual space is situated supero-medial to the mylohyoid muscle with its medial border demarcated by the genioglossus and hyoglossus muscles. It contains the sublingual glands, the lingual artery, vein and nerve plus hypoglossal and glossopharyngeal nerve branches and the deep portion of the submandibular gland (Fig. 6.8).

The submandibular or submaxillary space is situated inferolateral to the mylohyoid muscles and medial to the horizontal portion of the mandible. It contains the superficial portion of the submandibular gland, lymph nodes plus the facial artery and vein (Fig. 6.8).

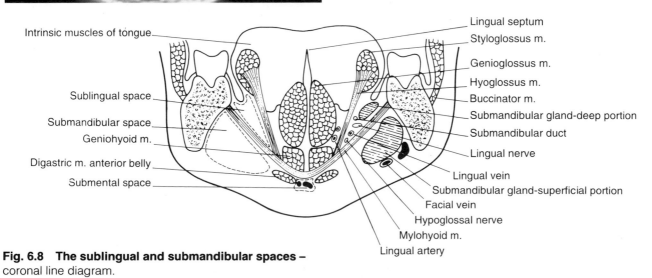

Fig. 6.8   The sublingual and submandibular spaces – coronal line diagram.

Another space, the submental space (Figs 6.7 and 6.8) is situated beneath the mylohyoid raphe between the anterior bellies of the digastric muscles. It contains fat and lymph nodes.

The hard palate is formed by components from three bones, the premaxilla, maxilla and palatine bones. It is covered by mucoperiosteum, and contains multiple minor salivary glands.

The gingivobuccal sulci are the paired upper and lower recesses formed between the cheeks and the alveolar ridges.

The retromolar trigone is a triangular area of mucosa behind the lower third molar tooth which extends on to the anterior aspect of the anterior tonsillar pillar. It is the commonest site of origin of squamous cell carcinoma within the pharynx.

Lymph drainage of the oropharynx, oral cavity and floor of the mouth

The lymphatic drainage of the oropharynx, oral cavity and floor of mouth is detailed in Fig. 6.9.

The soft palate drains into the jugulodigastric nodes but additional drainage pathways occur via the retropharyngeal, spinal accessory and submandibular nodes.

The anterior tonsillar pillars drain into the submandibular nodes and then into the deep cervical jugulodigastric nodes. The posterior tonsillar pillars drain into the jugulodigastric or submandibular nodes, with another possible drainage pathway via the spinal accessory nodes. The palatine tonsil drains directly into the jugulodigastric node.

The tongue has a complicated lymphatic drainage. Efferent lymphatics may drain to either side of the neck, particularly if a lesion is near the midline. The posterior third of the tongue drains to the deep cervical jugulodigastric nodes. The lateral and anterior tongue drain into the submandibular glands and from there into the deep cervical lymph node chain, particularly the jugulo-omohyoid node. Involvement of the retropharyngeal nodes has been demonstrated in patients with recurrent tongue tumours.

In the oral cavity, the gingivobuccal sulci, hard palate and retromolar trigone drain into submandibular and then deep cervical jugulodigastric lymph nodes. The lymph drainage of the floor of the mouth is to the submandibular and jugulodigastric nodes, although in a small number of patients drainage can also occur to the submental nodes.

The parapharyngeal spaces

The parapharyngeal spaces [1,2,7,8] are fat-filled triangular spaces situated lateral to the pharynx which extend from the base of skull down to the level of the hyoid bone (Fig. 6.10). The lateral boundary of each parapharyngeal space is formed by

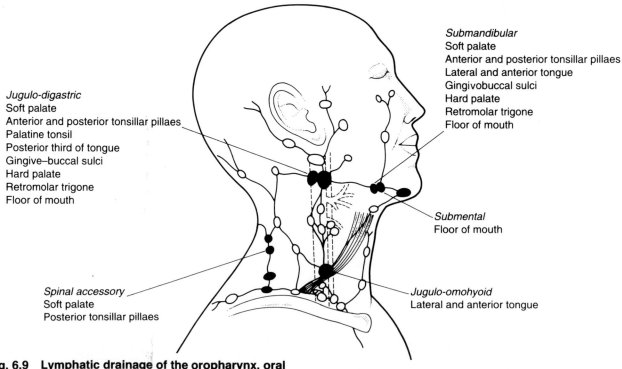

Jugulo-digastric
Soft palate
Anterior and posterior tonsillar pillaes
Palatine tonsil
Posterior third of tongue
Gingive–buccal sulci
Hard palate
Retromolar trigone
Floor of mouth

Submandibular
Soft palate
Anterior and posterior tonsillar pillaes
Lateral and anterior tongue
Gingivobuccal sulci
Hard palate
Retromolar trigone
Floor of mouth

Submental
Floor of mouth

Spinal accessory
Soft palate
Posterior tonsillar pillaes

Jugulo-omohyoid
Lateral and anterior tongue

Fig. 6.9 Lymphatic drainage of the oropharynx, oral cavity and floor of mouth.

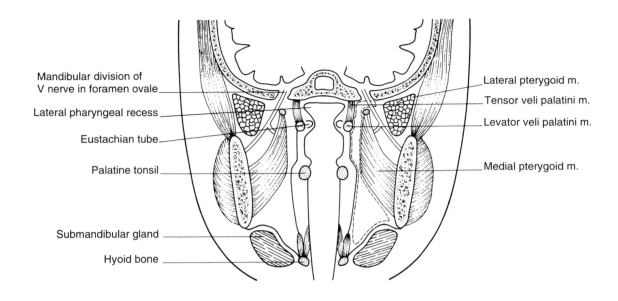

Mandibular division of V nerve in foramen ovale

Lateral pharyngeal recess

Eustachian tube

Palatine tonsil

Submandibular gland

Hyoid bone

Lateral pterygoid m.

Tensor veli palatini m.

Levator veli palatini m.

Medial pterygoid m.

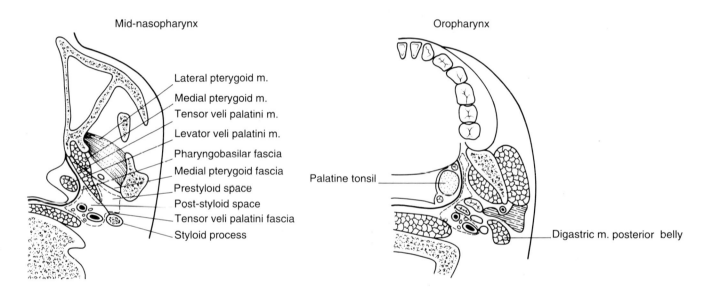

Mid-nasopharynx

Lateral pterygoid m.

Medial pterygoid m.

Tensor veli palatini m.

Levator veli palatini m.

Pharyngobasilar fascia

Medial pterygoid fascia

Prestyloid space

Post-styloid space

Tensor veli palatini fascia

Styloid process

Oropharynx

Palatine tonsil

Digastric m. posterior belly

Fig. 6.10 The parapharyngeal spaces. (a) Coronal plane. **(b)** Transaxial plane at the level of: **(i)** mid-maxillary antrum; **(ii)** hard palate; **(iii)** oropharynx. The parapharyngeal space is indicated by the dotted line.

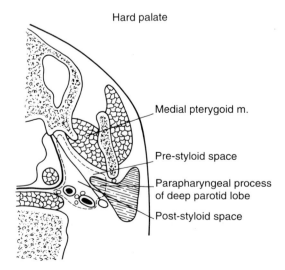

Hard palate

Medial pterygoid m.

Pre-styloid space

Parapharyngeal process of deep parotid lobe

Post-styloid space

the ascending ramus of the mandible, the insertion of the medial pterygoid muscles and the deep lobe of the parotid gland. The medial margin is formed by the naso- and oropharyngeal musculature. The anterior boundary of each parapharyngeal space is the pterygomandibular raphe, while supero-posteriorly its boundary is composed of the transverse process of C1, prevertebral muscles, the styloid process and the mastoid process.

Each parapharyngeal space is subdivided into an anterolateral prestyloid and a posteromedial retro-styloid compartment by a line joining the tensor veli palatini to the styloid process (Fig. 6.10). The presty-loid compartment contains the internal maxillary artery and branches of the mandibular division of the fifth cranial nerve (the inferior dental and auriculotemporal nerves) plus a small quantity of accessory salivary tissue. The retrostyloid compart-ment contains the internal carotid artery, the inter-nal jugular vein, the lateral retropharyngeal lymph nodes, the sympathetic chain and the ninth to twelfth cranial nerves.

The retropharyngeal space

The retropharyngeal space [1,2,8] is a potential midline space situated between the pharyngobasilar membrane and the prevertebral fascia, which is bounded laterally by the carotid sheath. Inferiorly, it is continuous with the retro-oesophageal space in the posterior mediastinum down to T6 level.

Pterygopalatine fossa

The boundaries of each pterygopalatine fossa [1,2,9] are the base of the skull superiorly, the posterior wall of the maxillary sinus anteriorly and the ptery-goid process posteriorly (Fig. 6.11). Medially its

boundary is the palatine bone. The pterygopalatine fossa contains the maxillary branch of the trigeminal nerve, the sphenopalatine ganglion and the terminal branch of the maxillary artery. The pterygopalatine fossa is important anatomically because it communi-cates with several other spaces and cavities, permit-ting tumour spread into these regions. There are connections with the infratemporal fossa via the pterygomaxillary fissure, with the nasal cavity via the sphenopalatine canal, with the oral cavity via the pterygopalatine canal, with the orbit via the inferior orbital fissure and with the cranial cavity via the pterygoid canal and foramen ovale.

Masticator space

The masticator space [1,2,9] contains the masseter muscle, the mandibular ramus and the medial and lateral pterygoid muscles. It is situated lateral to the parapharyngeal spaces and is involved by tumour spread from the nasopharynx after the parapharyn-geal space has been breached. The infratemporal fossa is that cranial portion of the masticator space immediately adjacent to the skull base.

THE NASOPHARYNX

Imaging modalities/techniques

Radiographs

If radiographs are required the examination should consist of:

1. a true lateral projection with the patient breathing through the nose, causing the soft palate to abut the tongue; and
2. a submento-vertical view with the mandible superimposed on the anterior table of the frontal bone.

The radiographs should be assessed for the presence of soft tissue masses, distortion of normal anatomy e.g. obliteration or displacement of the salpingo-pharyngeal fold, and bone abnormalities. The soft tissue anterior to the clivus and upper cervical vertebrae represents the roof and posterior wall of

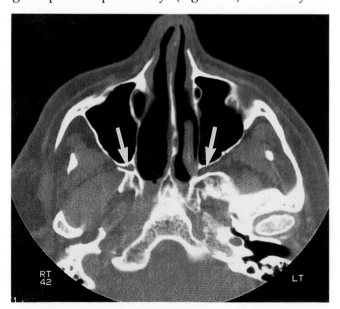

Fig. 6.11 The pterygopalatine fossae. Slightly asymmetrical transaxial CT scan with bone reformation demonstrating the pterygopalatine fossae (arrows) situated anterior to the pterygoid plate (on the right side) and the base of the skull (on the left side).

the nasopharynx. In relation to the clivus and C1 this soft tissue is of variable thickness, often with a blurred edge. It becomes more uniform between C2 and C4 where the thickness should be less than 0.5 cm. Below this level it becomes thicker, approaching 1 cm diameter in the post-cricoid region but should not exceed the antero-posterior diameter of the adjacent vertebral body. The thickness of the roof and posterior wall of the nasopharynx is greater in infancy and childhood due to lymphoid hyperplasia. A normal configuration may not be attained until the mid-twenties.

In general, patients with malignant tumours of the pharynx and oral cavity should have a chest radiograph, primarily to exclude co-existing lung tumours but also to detect thoracic metastases.

Computed tomography (CT) and magnetic resonance imaging (MRI)

Radiographs are of limited value in assessing the full extent of lesions of the nasopharynx and further imaging is usually required to provide greater anatomical detail and tissue discrimination. CT and MRI fulfil this role and have superseded conventional tomography and contrast studies. The relative advantages and limitations of these two techniques are discussed below, and the normal appearance of head and neck structures on spin-echo and gadolinium-enhanced MR images is detailed in Table 6.1.

Indications for CT and MRI examinations

The main indications are:

1. in patients with proven malignant tumours to assess extent of disease;
2. in patients presenting with signs and symptoms of malignancy in whom clinical/endoscopic examination is normal – the signs and symptoms include:
   (a) persistent unilateral serous otitis media
   (b) atypical facial pain
   (c) nasal obstruction or bleeding
   (d) cranial nerve palsies
   (e) cervical lympadenopathy
   Approximately 25% of patients with a single palpable malignant node in the mid or upper cervical chain will have a primary nasopharyngeal tumour;
3. to identify an appropriate biopsy site;
4. to evaluate benign tumours of the nasopharynx and parapharyngeal space;
5. to assess parapharyngeal masses of uncertain aetiology and to distinguish between parapharyngeal and parotid lesions.

CT technique

It may be helpful to examine the patient prior to performing the CT scan, although this is usually more beneficial in patients with oropharyngeal than

Table 6.1 Normal signal intensities of head and neck structures on conventional MR spin echo sequences

Tissue/organ	Signal intensity on T1-weighted image	Signal intensity on T2-weighted image	Enhancement with gadolinium DTPA on T1-weighted image
Pharyngeal mucosa	Intermediate	High	Yes
Pharyngobasilar fascia	Low	Low	No
Muscle	Intermediate to low	Low	No (slight)
Parotid gland	High	Becomes lower	Yes
Submandibular gland	High	Becomes lower	Yes
Pharyngeal and palatine tonsils	Low to intermediate	High	Yes
Soft palate	Intermediate	High	Yes
Lymph nodes	Intermediate	High	Yes
Fat	High	High	No
Air	Low	Low	—

nasopharyngeal lesions. Initial CT sections are obtained in the transaxial plane. The patient should lie supine with the neck comfortably extended and the head immobilized in a suitable support. Patients are asked not to move, cough or swallow during data collection. A lateral scout view should be obtained from which the examination is planned. The scan plane should be within plus or minus 10° from the infraorbital–meatal line, and a plane parallel to the hard palate provides a reproducible baseline.

Our practice in patients with proven malignant disease is to obtain an initial unenhanced scan with 1 cm thick contiguous sections taken from the roof of the orbit to the angle of the mandible. This scan is essentially an 'overview' and defines the area of primary interest through which an enhanced scan is obtained. If there is significant movement artefact then the enhanced sections should be obtained during suspended respiration. The enhanced scan should extend from the cavernous sinus to the soft palate. Sections should be contiguous with a thickness between 3 mm and 5 mm. Below the region of primary interest, 1 cm thick contiguous sections should be obtained to the level of the sternal notch to ensure that all lymph node drainage areas are included in the examination. Some authors advocate additional sections being obtained during manoeuvres such as a modified Valsalva to distend the fossae of Rosenmüller, to inflate the eustachian tube orifices and to see if mucosal 'masses' are mobile. In practice these are rarely necessary, particularly if the primary site of the tumour is known.

Patients with malignancy usually require a coronal scan to assess tumour extension to the skull base and cranial fossae. Ideally this scan should be obtained perpendicular to the hard palate or floor of the sphenoid sinus. Again 3–5 mm contiguous sections are required through the area of interest as assessed from the axial scan.

Prospectively images should be obtained using the soft tissue or 'standard' reconstruction algorithm. Retrospective reconstruction from the raw data is used to improve bone detail, especially of the skull base, and spatial resolution in particular areas of interest by using a smaller field of view.

The above technique can be modified in patients with benign disease where an enhanced scan through a limited area may suffice. MRI may be a more appropriate examination in these patients.

MRI technique

While CT is still the most widely used cross-sectional imaging technique for evaluation of the naso- and oropharynx, MRI will provide the same anatomical information and has certain advantages:

1. no ionizing radiation;
2. improved tissue discrimination;
3. a multiplanar facility;
4. intravenous contrast often unnecessary, but if required, safer than iodine compounds in the volumes used;
5. improved specificity in distinguishing between paranasal sinus obstruction or invasion by tumour;
6. improved specificity in distinguishing between residual/recurrent tumour and fibrosis;
7. MR angiography will provide information about the relationship of the major vascular structures to a mass lesion.

The nasopharynx can usually be imaged adequately in the standard head coil; a combined head and neck coil will allow simultaneous assessment of the cervical lymph nodes. When the area of maximum interest is centred in the receiver and the head is secured, patients are asked to remain as still as possible and keep swallowing to a minimum. Multislice T1- and T2-weighted images, with a slice thickness in the order of 5 mm, should be obtained in at least two planes through the disease site. The use of presaturation and flow compensation techniques will reduce some of the motion artefact, although they are more frequently used when imaging the oral cavity or oropharynx.

The choice of plane of section and pulse sequences will vary according to the site of disease and machine capability. In the first instance it is appropriate to obtain T1-weighted transaxial images to demonstrate the overall anatomy and symmetry of the deep tissue planes of the nasopharynx and parapharyngeal region. Retropharyngeal nodes are best demonstrated in this plane. T2-weighted images are obtained in either the coronal or sagittal plane. Coronal plane images should be obtained when a lesion involves the nasopharyngeal roof and to assess changes around the skull base, in the cavernous sinuses and middle cranial fossae. Sagittal images are best for assessing changes in relation to the basisphenoid and clivus. Craniocaudal extent of disease is best assessed on coronal or sagittal images and lateral extent on transaxial or coronal images.

CT versus MRI

CT is cheaper, easier to use and less motion-sensitive than MRI. It provides better bone detail and in most circumstances a higher spatial resolution. There is evidence that CT is better at assessing cervical lymph node metastases and that dynamic contrast studies provide greater tissue specificity. The decision on which technique to use depends on a number of factors, including the patient's clinical status, the current clinical problem, the available equipment and machine time and the requirement to limit radiation exposure/dose.

In relation to head and neck cancer CT is often performed as the initial examination in patients with sinonasal and oropharyngeal tumours. In the nasopharynx MRI is probably the 'first choice' examination as the improved tissue discrimination defines the extent of deep infiltration of tumour more accurately. A complete evaluation of the patient may require both CT and MRI examinations.

Intravenous contrast administration

CT

The basic indications for and aims of using intravenous CT contrast agents are:

1. to try and improve the staging of primary tumours;
2. to assess the relationship of the major vascular structures to the primary tumour;
3. to characterize enlarged cervical lymph nodes;
4. to characterize parapharyngeal masses, which can encase or displace adjacent vessels in a characteristic fashion;
5. to differentiate tumour in the paranasal sinuses from obstruction secondary to tumour within the nasopharynx and nasal cavity;
6. to differentiate between residual/recurrent tumour and post-treatment fibrosis.

In practice the majority of patients with lesions in the nasopharynx and parapharyngeal space require an enhanced CT scan. The aims described above require the iodine to be in the vascular phase during the examination. In order to achieve this the consensus of opinion is that a bolus infusion technique is optimal. However the iodine concentration, volume and osmolality of the contrast medium used is influenced by a number of factors, not least of which is cost.

MRI

Intravenous gadolinium-based contrast agents cause T1 shortening and increased signal intensity of lesions on T1-weighted images. The relatively short imaging time required for a T1-weighted sequence (compared to a T2-weighted sequence) reduces motion artefact and further improves lesion conspicuity. The use of gadolinium therefore improves tumour delineation.

The indications for intravenous MR contrast are:

1. to define the tumour margin and differentiate tumour from surrounding oedema;
2. to assess deep extension of tumour;
3. to identify perineural tumour infiltration;
4. to identify leptomeningeal tumour involvement;
5. to differentiate tumour within the sinuses from inspissated secretions;
6. to help differentiate residual or recurrent tumour from post-treatment fibrosis.

Evaluation of CT and MR images

As well as the nasopharynx, adjacent structures must be evaluated. These include the skull base, the pterygopalatine fossa, the parapharyngeal and masticator spaces, the cervical spine and the base of the tongue. All images should be interrogated on a diagnostic console at bone and soft tissue settings. It is unwise to issue a report on review of hard copy images alone.

Angiography

Angiography has a role in the diagnosis and management of vascular tumours within the nasopharynx.

Malignant tumours are usually poorly vascularized with no distinctive angiographic features, but some benign tumours, in particular juvenile angiofibromas, have a characteristic appearance. These tumours spread outside the confines of the nasopharynx and such extension can be delineated at angiography. Embolization techniques are used as an adjunct to other forms of treatment.

Benign lesions of the nasopharynx

Inflammatory lesions

Infections

Nasopharyngeal infections rarely require radiological investigation except when complications such as parapharyngeal or retropharyngeal abscesses supervene (see below).

Other inflammatory lesions

The most common inflammatory lesions are usually related to the adenoids and present as mass lesions (see below). Occasionally, aggressive inflammatory lesions occur with patterns of spread similar to malignancy, including bone destruction. The CT and MRI appearances are indistinguishable from malignancy and diagnosis is usually based on the clinical history, examination and biopsy. Malignant external otitis media, an infection often occurring in patients with diabetes mellitus or in those who are debilitated, can spread to involve the parapharyngeal space and skull base and simulate malignancy.

Postoperative changes and immediate postbiopsy changes can also produce inflammatory changes which may simulate malignancy, and therefore diagnostic imaging should be performed before or at least 7–10 days after these procedures.

Mass lesions of the nasopharynx

These include lymphoid hyperplasia, tumours and cysts. Symptoms are usually the result of local mass effect producing nasal obliteration, nasal speech or obstruction to the eustachian tube, which in turn produces serous otitis media, and occasionally epistaxis if the lesion is vascular.

Benign tumours are uncommon and consist of those presenting as pedunculated lesions in the nasopharynx e.g. choanal polyps, inflammatory polyps, inverting papillomas and juvenile angiofibroma and those arising deep to the mucosa or in the parapharyngeal space which bulge into the airway.

Lymphoid hyperplasia

The nasopharynx has abundant lymphatic tissue (adenoids) which can become hyperplastic in children and adolescents, occasionally producing hypersomnolence (Pickwickian syndrome). Involution occurs in late adolescence but may not be complete until the mid-twenties. Lymphoid hyperplasia also occurs in response to upper respiratory tract infections, including infectious mononucleosis. The majority of patients do not require imaging but lymphoid hyperplasia is seen as an incidental finding on CT/MRI scans in children and young adults and cross-sectional imaging will occasionally be indicated if the symptoms simulate tumour.

Lymphoid hyperplasia usually appears as a symmetrical mass about the midline which is restricted to the mucosal side of the pharyngobasilar fascia. On CT, normal adenoids are usually of slightly lower attenuation than muscle and show little or no

enhancement with contrast [10] (Fig. 6.12), although there may be enhancement of the lamina propria. On MRI, lymphoid tissue is of high signal intensity especially on T2-weighted pulse sequences (Figs 6.12 and 6.13).

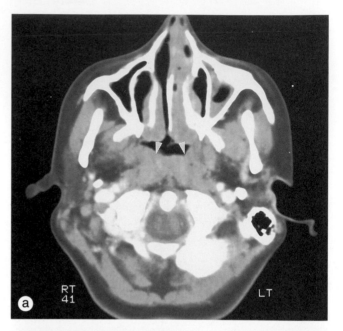

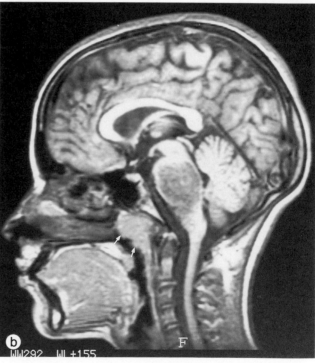

Fig. 6.12 Nasopharyngeal lymphoid hyperplasia.
(**a**) Transaxial contrast-enhanced CT scan demonstrating nasopharyngeal lymphoid hyperplasia (arrowheads) in a 20-year-old. (**b**) Sagittal T1-weighted gradient echo image demonstrating lymphoid hyperplasia (arrows) in a 13-year-old. The patient has a malignant astrocytoma of the pons.

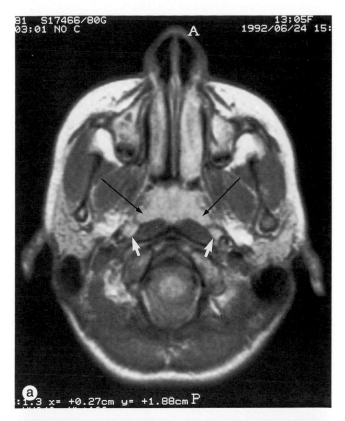

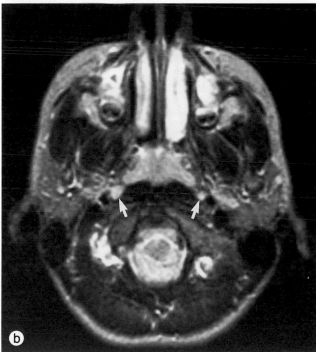

Fig. 6.13 Lymphoid hyperplasia obliterating the lateral pharyngeal recesses. First (**a**) and second (**b**) echo T2-weighted transaxial images demonstrating lymphoid hyperplasia extending into the lateral pharyngeal recesses (black arrows in (**a**)). There are also small bilateral retropharyngeal nodes (arrows).

Localized hypertrophy of lymphoid tissue can occur around the eustachian tube resulting in obliteration of the fossae of Rosenmüller and retropharyngeal nodes up to 2 cm may be found (Fig. 6.13). These two features may lead to some difficulty in diagnosis.

Choanal polyps

These are hyperplastic inflammatory polyps arising from the maxillary sinus which protrude into the nasal cavity and thence into the nasopharynx. They usually occur in non-atopic adolescents and young adults where, despite their size, they are often asymptomatic and discovered incidentally (Chapter 5).

Inflammatory polyps can arise from the roof of the nasopharynx and present as pedunculated masses. Without an appropriate clinical history and associated findings biopsy is required for diagnosis.

Juvenile angiofibroma

This is the commonest benign tumour of the nasopharynx. It is an enlarging vascular tumour thought to arise from the mucosa overlying the body of the sphenoid or roof of the posterior nasal cavity. Some authors believe it to originate in the pterygopalatine fossa gaining access to the nasopharynx through the sphenopalatine foramen. The histogenesis is uncertain and like haemangioma it may be a hamartoma of the nasal erectile tissue. Microscopically it consists of small vascular spaces within a poorly cellular fibrous tissue.

Juvenile angiofibroma classically occurs in teenage males (range 7–32 years) who present with symptoms of nasal obstruction and spontaneous epistaxis and have a visible mass on examination. Cases have been reported in females and the tumour is known to undergo spontaneous regression.

The tumour grows slowly through natural foraminae and fissures producing pressure bone erosion. Anteriorly tumour extends into the nasal cavity and can reach the anterior nares. Spread occurs through the ostia into the maxillary antrum, the sphenoid and ethmoid sinuses. Tumour in the pterygopalatine fossa produces a characteristic expansion of the fossa with anterior displacement of the posterior wall of the maxillary antrum and posterior displacement of the pterygoid plates [11]. Bone is thinned but not irregularly eroded as with a malignant process. From the pterygopalatine fossa tumour can extend either into the infratemporal fossa and on to involve the parapharyngeal space, or it can extend into the orbit via the inferior orbital fissure and thence via the

157

superior orbital fissure into the middle cranial fossa. Initially intracranial tumour is extradural but invasion of the dura, meninges and brain can occur. Tumour within the sphenoid sinus may erode through into the cavernous sinus.

If the diagnosis is suspected then CT is usually the initial investigation, both to corroborate the diagnosis and to assess the extent of tumour. The CT appearances reflect the pattern of growth described above and are illustrated in Fig. 6.14. In most cases the tumour appears to be centred on the pterygopalatine fossa. An enhanced scan should always be obtained, on which the tumour will demonstrate intense staining. The use of contrast enables a distinction to be made between invaded and obstructed sinuses and provides a better discrimination between normal and abnormal soft tissue structures, especially when there is intracranial extension. Particular attention should be paid to the cavernous sinus if there is tumour within the sphenoid sinus. Because the pattern of intracranial extension may be difficult to appreciate on the transaxial images,

coronal images should be obtained. On MRI, multiple small signal voids reflect the vascularity of the lesion (Fig. 6.14(c)).

Angiography is undertaken to identify the blood supply prior to surgery and for therapeutic embolization. Occasionally it is used as a diagnostic procedure to distinguish between an angiofibroma and the rarer angioma.

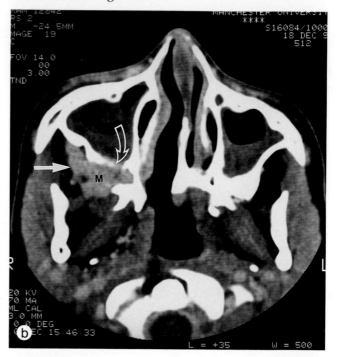

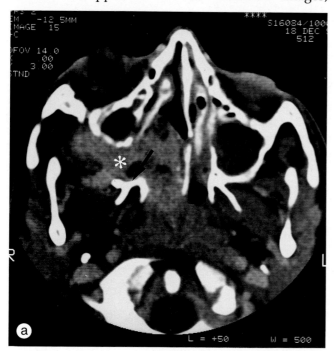

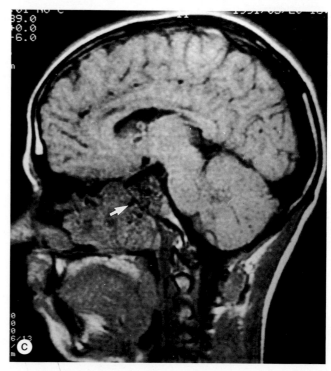

Fig. 6.14 Juvenile angiofibroma. Post-contrast transaxial CT scans at the level of the floor of the orbit (**a**) and mid-maxillary antrum (**b**). There is a right-sided enhancing soft tissue mass (M) involving the nasopharynx, right nasal cavity and pterygopalatine fossa (asterisk) and producing pressure erosion of the posterior wall of the right maxillary antrum (open curved arrow). The pterygoid plate shows characteristic posterior bowing (straight black arrow). Disease is extending into the right infratemporal fossa (white arrow). (**c**) Sagittal T1-weighted MR image demonstrating multiple signal voids (arrow) within the tumour mass.

Thornwaldt's cyst

These are uncommon and thought to represent failure of notochord regression in the nasopharynx. The potential space or bursa lined with respiratory epithelium can give rise to a true cyst in the midline or to either side of the midline. They may be asymptomatic or produce a postnasal discharge. CT characteristics are of a cystic mass with no involvement of the underlying bone, thereby excluding a meningocele. On MR, the proteinaceous contents of a Thornwaldt's cyst are of very high signal intensity on T2-weighted images.

Rarely, craniopharyngiomas, sphenoid sinus mucoceles or pituitary adenoma will present as a nasopharyngeal mass.

Parapharyngeal space masses [12]

As indicated these can present as a mass lesion bulging into the nasopharynx. Primary tumours in this region account for approximately 0.5% of head and neck cancer and high quality CT or MR scans will demonstrate the origin and extent of the lesion and predict the diagnosis in 80–90% of cases. This prediction is based on the site of the lesion and its characteristics on unenhanced and enhanced scans.

Primary tumours in the prestyloid space usually arise in accessory salivary glands, the majority being benign pleomorphic adenomas. Patients may notice a bulge on the soft palate which is displaced inferomedially or be aware of a mass in the parotid region. The masses are usually painless. CT appearances are of a rounded, well-circumscribed soft tissue mass isodense with muscle which may or may not enhance with contrast. Low attenuation areas within the lesion may be seen, representing mucoid impaction or necrosis. Calcification is uncommon. The mass will displace the internal carotid artery and jugular vein postero-laterally and it is important to distinguish such lesions from those arising in the deep lobe of the parotid gland which are secondarily involving the parapharyngeal space, as the surgical approach is different. Distinction is based on the presence or absence of the fat plane between the deep lobe of the parotid and the mass, this plane being absent in lesions arising from the deep lobe (Fig. 6.15).

Other lesions arising in the prestyloid compartment are rare and include lipomas, liposarcomas and branchial cleft cysts.

Neurogenic tumours (schwannoma, neurofibroma and paraganglioma) form the bulk of the primary lesions arising in the poststyloid compartment and account for approximately 30–40% of primary para-

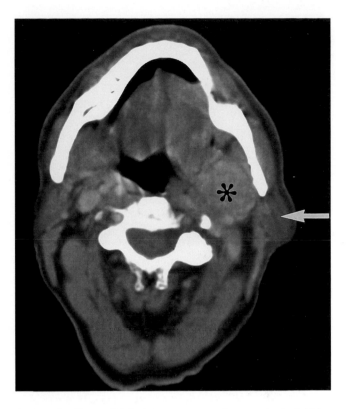

Fig. 6.15 Prestyloid space tumour. A mass (asterisk) in the prestyloid space is shown to be separate from the parotid (arrow) because of the presence of a fat plane between them.

pharyngeal lesions [13]. The commonest lesions are schwannomas, which usually arise from the tenth cranial nerve or the cervical sympathetic trunk. Paragangliomas arising from neural crest cells include the glomus vagale, glomus jugulare and carotid body tumours: the latter two secondarily involve the parapharyngeal space. Neurofibromas are frequently associated with neurofibromatosis.

Schwannomas (Fig. 6.16) and neurofibromas can arise from the cranial nerves IX–XII or the cervical sympathetic chain. The symptoms are referable to the nerve involved and pressure on adjacent structures. These lesions may cause some displacement of the carotid artery medially and the jugular vein laterally. On CT schwannomas are usually rounded, well-circumscribed lesions exhibiting a heterogeneous pattern of enhancement.

Paragangliomas often present with tinnitus and are also well-circumscribed but lobulated lesions obliterating the interfaces between the carotid artery and jugular vein. Approximately 75% demonstrate uniform enhancement on CT, the other 25% heterogeneous enhancement (Chapter 3).

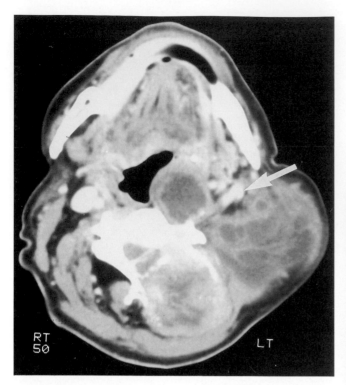

Fig. 6.16 Poststyloid space schwannoma. Huge, heterogeneously enhancing schwannoma with a component indenting the left oropharynx and, in this case, laterally displacing both the carotid artery and internal jugular vein (arrow).

An enlarged retropharyngeal node (>1 cm) is often indistinguishable from a parapharyngeal mass (Fig. 6.4). These nodes lie medial to the carotid sheath and hence displace it postero-laterally. The nodes may be involved by lymphoma or metastatic tumour, or be inflammatory. Enhancement characteristics do not help to differentiate between benign and malignant lesions. Nodes may or may not enhance. Those involved with lymphoma and non-necrotic inflammatory nodes often exhibit homogeneous enhancement. Ring enhancement is seen in both necrotic and non-necrotic metastatic nodes and inflammatory nodes.

The parapharyngeal space may be involved by lesions arising from the skull base or CNS structures e.g. meningioma, glomus jugulare, neurogenic tumours, chordomas and epidermoid/dermoid cysts (see Chapter 3).

If there is any doubt about the vascularity of a lesion in the parapharyngeal space and surgery is contemplated then preoperative angiography should be peformed.

Other pathology affecting the parapharyngeal space includes jugular vein thrombosis, the most common cause of which is an indwelling central venous catheter. Other recognized causes include

local surgery, infection and intravenous drug abuse. More rarely, the jugular vein may be compressed by neck tumours and there is also an association with malignant coagulopathy.

On CT, there is increased size of the affected vein and the thrombus is seen as a low attenuation non-enhancing filling defect [14]. There may be marginal enhancement due to blood flow around the periphery of the clot and through the vasa vasorum of the vessel wall (Fig. 6.17). Secondary reactive soft tissue swelling may be seen.

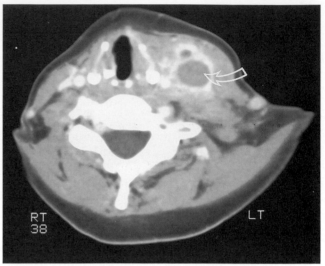

Fig. 6.17 Internal jugular vein thrombosis. Transaxial contrast-enhanced CT scan demonstrating a left internal jugular vein thrombosis (arrow) with involvement of a peripheral venous branch. Marginal enhancement is seen around the thrombosed vein.

On MRI intravascular thrombus is indicated by signal within the jugular vein on conventional spin echo sequences and by intermediate signal intensity (as opposed to high signal intensity in flowing blood) on gradient-recalled echo sequences.

Malignant tumours of the pharynx [18]

Aetiology

Both nasopharyngeal and oropharyngeal/oral cavity carcinomas are commoner in males, with male:female ratios of 3:2 in the nasopharynx and 4:1 in the oropharynx and oral cavity. Factors implicated in the aetiology of pharyngeal tumours include alcohol, tobacco, chronic infection, poverty and marijuana abuse.

Undifferentiated nasopharyngeal carcinoma, which is rare in most parts of the world, is common in south-east China, Greenland and north and central Africa and is associated with Epstein–Barr virus infection. The epithelial tumour cells carry the Epstein–Barr virus DNA and express Epstein–Barr nuclear antigen, with affected individuals having high titres of antibodies to Epstein–Barr viral antigens. There is also an association with tissue antigens e.g. HLA-A$_2$, indicating a genetic factor for tumour development. Further support for a genetic aetiology comes from the study of Chinese domiciled outside of China who have an incidence of nasopharyngeal carcinoma approximately five to seven times that of the native population.

Oral cavity tumours are associated with liver cirrhosis, possibly because of vitamin deficiency, and there is also an association with Plummer–Vinson syndrome.

All patients with pharyngeal malignancies are at increased risk of developing oesophageal carcinoma and malignant tumours of the salivary glands and breast. There is also a well-recognized increased risk of primary lung tumours, which occur in 6–10% of patients.

Pathology

Pharyngeal tumours are most commonly squamous cell carcinomas, which account for 80% of nasopharyngeal lesions and 70% of oropharyngeal and oral cavity lesions. Squamous cell carcinoma may be further subdivided into keratinizing and non-keratinizing subgroups. Transitional cell carcinomas are included in the non-keratinizing subgroup, and so-called 'lymphoepitheliomas' actually represent poorly differentiated squamous cell carcinomas interspersed with normal lymphocytes. Other recognized squamous cell carcinoma variants include spindle cell carcinoma, verrucous carcinoma and Schmincke tumours. Pharyngeal adenocarcinoma is unusual and typically arises in the numerous glands of the soft palate. Undifferentiated carcinoma accounts for approximately 10% of nasopharyngeal lesions.

Twenty-five per cent of oropharyngeal and oral cavity tumours are non-Hodgkin's lymphomas and lymphoma accounts for 15% of nasopharyngeal lesions. Other rare tumours which may occur in the pharynx are plasmacytomas, melanomas, small-cell tumours, and rhabdomyosarcoma, which affects children. Esthesioneuroblastomas arise in the nasal cavity and then spread to the nasopharynx.

The role of imaging in pharyngeal malignancies

Accurate delineation of the primary tumour together with the identification of lymph node metastases strongly influences clinical management and cross-sectional imaging is of undoubted value in all but the most superficial mucosal tumours. It has been shown to accurately assess the size of the primary lesion, and to identify spread into adjacent organs, tissues and spaces.

On CT the primary lesion is typically isodense with muscle but demonstrates enhancement after intravenous contrast administration. On MR T1-weighted images squamous cell tumours are of low signal intensity or isointense with muscle. They demonstrate high signal intensity on T2-weighted images and enhancement after intravenous paramagnetic contrast administration.

The use of gadolinium has been shown to improve tumour/muscle contrast and to help differentiate between solid tumour, central necrosis and surrounding oedematous change [16, 17].

MRI is also valuable in identifying perineural tumour spread, when there may be thickening of involved nerves, expansion of skull base foramina, abnormal signal intensity within nerve ganglia and atrophy of muscles within the distribution of an involved nerve [18].

The incidence of metastases depends on the degree of differentiation of the tumour, the size of the primary lesion, vascular space invasion and the number of capillary lymphatics in the affected tissue. Lymphatic spread to outlying and then regional lymph nodes occurs in approximately 60% of patients at presentation, although incidence varies depending on the site of the primary tumour. For example only 10–15% of patients with confined clinical Stage T1 lesions of the soft palate or tongue have nodal metastases whereas over 90% of nasopharyngeal Stage T1 tumours have lymph node involvement. Extracapsular nodal spread is a poor prognostic indicator in that patients have a shorter disease-free interval between treatment and recurrence compared to those with tumour confined within the nodes.

Assessment of cervical lymph nodes can be made on CT and MRI. This is particularly important in the case of clinically impalpable groups such as the superior deep cervical nodes.

Since neither CT density changes, nor MRI tissue characterization permit reliable detection of lymph node tumour spread, identification of nodal involvement primarily depends on size. Nodes measuring greater than 1 cm in their maximum short-axis diameter on transaxial sections are considered to have a high probability of tumour involvement. Lymph nodes, even if smaller than 1 cm, which have a necrotic or cystic centre (Fig. 6.18) are also considered to be infiltrated by tumour [19]. Contrast-enhanced CT has been shown to be superior to both

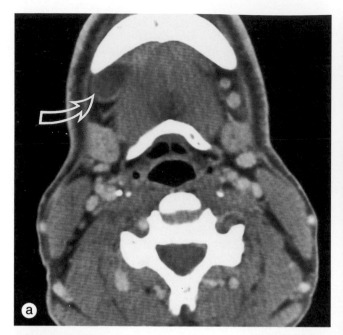

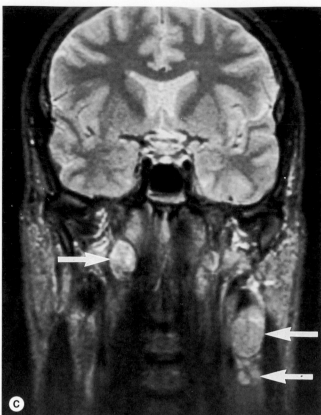

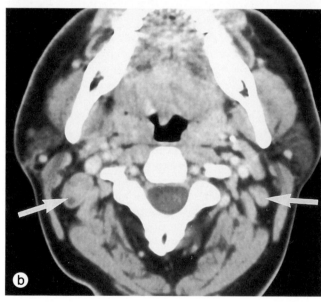

Fig. 6.18 Cystic and necrotic malignant lymph nodes.
(**a**) Transaxial CT scan demonstrating an involved cystic right submandibular node (curved arrow) in a patient with oropharyngeal squamous carcinoma. (**b**) Transaxial CT scan demonstrating partially necrotic deep cervical nodes (arrows). (**c**) Coronal T2-weighted MR image demonstrating markedly enlarged cervical lymph nodes (arrows) which are of heterogeneous high signal intensity.

unenhanced and enhanced MRI in the detection of central nodal necrosis and extracapsular nodal spread [20].

After intravenous contrast administration, ring enhancement may be observed within involved lymph nodes, but this is non-specific since it may also occur in inflammatory nodes.

Distant metastases most commonly occur to the lung, mediastinal nodes or bone and are seen within 3 years of diagnosis. Therefore a chest X-ray is usually performed during staging to assess the patient for the presence of metastases or any coexistent lung primary.

Treatment of pharyngeal malignancy

Nasopharyngeal tumours are usually treated by radiotherapy. Treatment of oropharyngeal and oral cavity lesions is by surgery, radiotherapy or combined surgery and radiotherapy. Head and neck tumours are often resistant to chemotherapy.

Malignant tumours of the nasopharynx

Nasopharyngeal carcinoma

Clinical features

Nasopharyngeal carcinoma can occur at any age, although there is a bimodal peak for undifferentiated tumours between 10–20 years and 45–55 years. Presenting signs and symptoms relate to the local and distant spread of the tumour.

The commonest symptoms are otic and include tinnitus and loss of hearing, which result from obstruction to, or direct extension along, the eustachian tube and the development of serous otitis media. Anterior extension of tumour produces nasal obstruction with resulting 'stuffiness', mucoid or mucopurulent discharge and epistaxis. Symptoms related to the paranasal sinuses result from obstruction of their ostia. Neurological symptoms include:

1. headache, usually secondary to intracranial extension, either directly or via the basal foramina;
2. ophthalmoplegia as a result of direct involvement of the orbit or indirectly of the cavernous and sphenoid sinuses;
3. involvement of the fifth cranial nerve.

Fifth cranial nerve involvement is reportedly common and results from either intra- or extracranial disease. The sites of involvement are usually:

1. the prestyloid compartment of the parapharyngeal space, with consequent pain in the distribution of the third division and atrophy of the muscles of mastication;
2. the pterygopalatine and infratemporal fossae by extension of tumour from the parapharyngeal space;
3. the foramen ovale, which can be involved by intracranial, extradural extension of tumour along the carotid arteries or by extension of tumour from the parapharyngeal space;
4. the cavernous sinus, which produces paraesthesia and pain in the distribution of the second and third divisions.

Involvement of the poststyloid compartment of the parapharyngeal space, either by direct spread or by lymphadenopathy can result in palsies of cranial nerves IX, X, XI and XII and involvement of the cervical sympathetic chain. Extension can occur from this site by perineural or perivascular pathways.

Nasopharyngeal tumours are staged according to the TNM classification (Table 6.2). The TNM staging classification for nodal and distant metastases is detailed in Table 6.3.

Table 6.2   TNM staging classification: nasopharynx

T – **Primary tumour**

TX	Primary tumour cannot be assessed
T0	No evidence of primary tumour
Tis	Carcinoma *in situ*
T1	Tumour limited to one subsite of nasopharynx
T2	Tumour invades more than one subsite of nasopharynx
T3	Tumour invades nasal cavity and/or oropharynx
T4	Tumour invades skull and/or cranial nerve(s)

CT and MRI appearances

The CT and MRI appearances [21] of nasopharyngeal carcinoma reflect the growth patterns at the primary site and mode of spread (Figs 6.19–6.21).

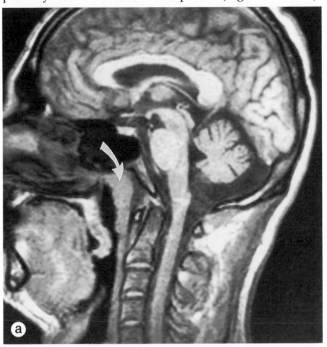
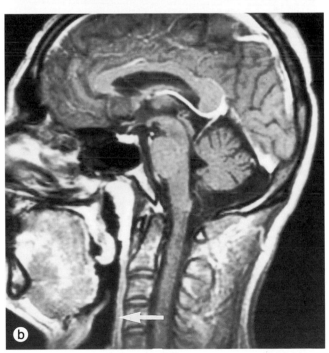

Fig. 6.19   Nasopharyngeal carcinoma. T1-weighted precontrast (**a**) and T1-weighted postcontrast (**b**) sagittal MR images demonstrating an infiltrative enhancing tumour (white curved arrows) in the supero-posterior nasopharynx. Note destruction of the posterior pharyngeal muscle/raphe, which is preserved more inferiorly (straight arrow).

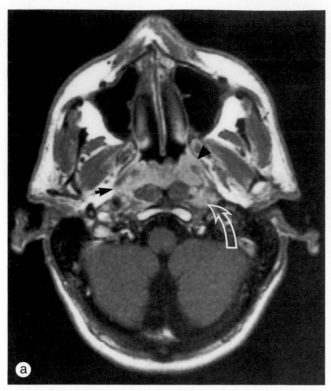 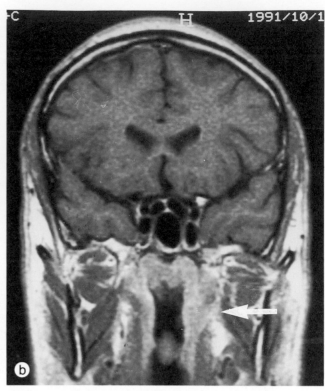

Fig. 6.20 Nasopharyngeal carcinoma. Transaxial (**a**) and coronal (**b**) contrast-enhanced T1-weighted MR images showing a nasopharyngeal tumour which demonstrates the following:

1. extension through the right pharyngobasilar fascia (small black arrow in (**a**)) into the right parapharyngeal space with involvement of the levator veli palatini muscle;
2. preservation of the left pharyngobasilar fascia at the same level but anterior extension around the left torus tubarius into the eustachian orifice and probably through the foramen of Morgagni (black arrowhead in (**a**));
3. extension into the left parapharyngeal space at a lower level (white arrow in (**b**));
4. retropharyngeal lymphadenopathy (curved arrow in (**a**));
5. an intact skull base on the coronal view.

Table 6.3   TNM staging classification of nodal disease and metastases in head and neck tumours

N/pN – Regional lymph nodes

N/pNX	Regional lymph nodes cannot be assessed
N/pN0	No regional lymph node metastasis
N/pN1	Metastasis in a single ipsilateral lymph node, 3 cm or less in greatest dimension
N/pN2	Metastasis in a single ipsilateral lymph node, more than 3 cm but not more than 6 cm in greatest dimension; **or** in multiple ipsilateral lymph nodes, none more than 6 cm in greatest dimension; **or** in bilateral or contralateral lymph nodes, none more than 6 cm in greatest dimension
N/pN2a	Metastasis in a single ipsilateral lymph node, more than 3 cm but not more than 6 cm in greatest dimension
N/pN2b	Metastasis in multiple ipsilateral lymph nodes, none more than 6 cm in greatest dimension
N/pN2c	Metastasis in bilateral or contralateral lymph nodes, none more than 6 cm in greatest dimension
N/pN3	Metastasis in a lymph node, more than 6 cm in greatest dimension

M – Metastasis

MX	Presence of distant metastasis cannot be assessed
M0	No distant metastasis
M1	Distant metastasis

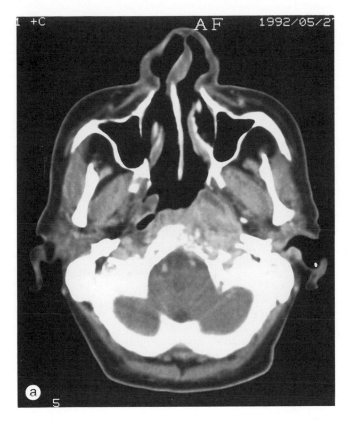

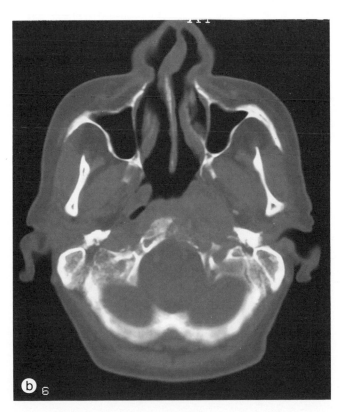

The tumours may be exophytic, ulcerative, infiltrative or any combination of these. They arise most frequently in the fossa of Rosenmüller or the region of the torus tubarius, less frequently on the roof, posterior wall and the nasopharyngeal surface of the soft palate. It may not be possible to identify the site of origin with large tumours or occult infiltrative lesions, when there may be no mucosal disruption.

Tumour extends along pathways of least resistance, with certain structures acting as barriers and modifying progress. Cartilage is relatively resistant to tumour with the result that tumour tends to extend around the cartilaginous portion of the eustachian tube and the cartilage in the foramen lacerum rather than destroying it. The relative resistance of the pharyngobasilar fascia makes it possible for tumour to spread from the skull base to the oropharynx, or *vice versa*, within the fascia, causing asymmetry, often subtle, of the pharyngeal walls on the scans and occlusion of the torus tubarius. Transgression across the pharyngobasilar fascia usually indicates malignancy, although rarely an aggressive inflammatory lesion will do this.

Extension into the parapharyngeal space is the most reliable indicator of tumour. Spread is either directly through the pharyngobasilar fascia or follows the eustachian tube through the sinus of Morgagni. Once in the parapharyngeal space tumour can extend to the skull base, producing bone erosion, and thence through the foramina into the skull (Fig. 6.22). Involvement of the levator veli palatini as it passes with the eustachian tube through the sinus of Morgagni or the tensor veli palatini, which lies external to the pharyngobasilar fascia, may occur with consequent spread along the muscle producing asymmetry of the pharyngeal walls and bone erosion at their sites of origin and insertion. Direct spread to the poststyloid compartment can result in involvement of the carotid sheath, the internal carotid artery lying less than half a centimetre from the fossa of Rosenmüller. Tumour can follow the internal carotid artery through the carotid canal into the cavernous sinus, or follow the jugular vein into the jugular foramen. This type of tumour extension may also cause bone erosion as can direct superior extension of tumour from the primary site.

Fig. 6.21 Nasopharyngeal carcinoma involving the skull base. Transaxial post-contrast CT scan (**a**) and a reconstruction using a bone algorithm (**b**), showing a large left nasopharyngeal mass obliterating the lateral recess and destroying the left side of the clivus.

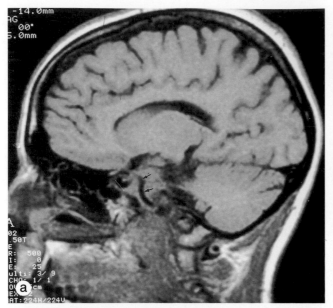

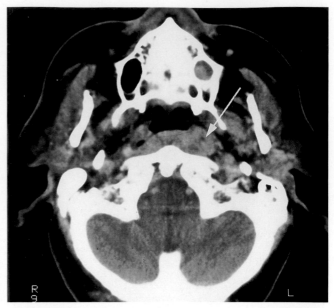

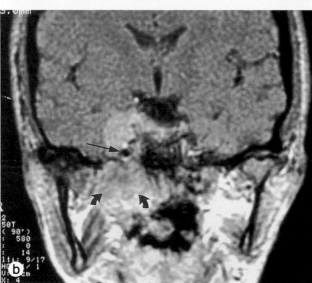

Fig. 6.22 Intracranial extension of nasopharyngeal carcinoma. (a) Sagittal T1-weighted image demonstrating the signal void of the normal internal carotid artery (arrows) encased by intermediate signal tumour. **(b)** Coronal T1-weighted image following gadolinium DTPA. Enhancing tumour is seen in the nasopharynx (curved arrows) extending upwards through the skull base foramina into the right parasellar region. Note the position of the internal carotid artery flow void (long arrow). (Reproduced by courtesy of Dr T Jaspan.)

In addition to the pathways of direct spread described above there may be lymphatic and haematogenous metastases.

In patients presenting with malignant neck nodes from an unknown primary, CT or MRI may be of value in identifying the site of the primary lesion (Fig. 6.23).

Fig. 6.23 Malignant lymphadenopathy from a clinically occult nasopharyngeal primary. The patient presented with a malignant right upper deep cervical node (Fig. 6.18(**b**)). At that time ENT examination was normal. Two months later transaxial CT demonstrated a left nasopharyngeal mass (arrow).

Lymphoma

In addition to cervical lymph nodes, lymphoma of the head and neck can involve Waldeyer's ring, thyroid, salivary glands, base of the tongue, ear, nasal cavity and paranasal sinuses, orbit and skin. These sites account for approximately 15% of extra-nodal non-Hodgkin's lymphoma, Waldeyer's ring being the commonest site of involvement. Waldeyer's ring consists of the lymphoid tissue of the nasopharynx, the base of the tongue, palatine tonsil, soft palate and posterior oropharyngeal wall. Within Waldeyer's ring the palatine tonsil is the commonest site of involvement, followed by the nasopharynx and the base of the tongue. Associated cervical lympadenopathy is present in approximately 50% of patients, retroperitoneal adenopathy in 30% and gastrointestinal involvement in 20%.

Clinical presentation and radiology

Lymphoma tends to present as an exophytic, non-ulcerating lesion without invasion of adjacent tissues. Lesions can be solitary or multiple. The clinical symptoms are often similar to those of nasopharyngeal carcinoma and the diagnosis is made histologically following biopsy. The radiological findings are indistinguishable from other exophytic tumours (Fig. 6.24), but a characteristic feature is circumferential involvement of Waldeyer's ring.

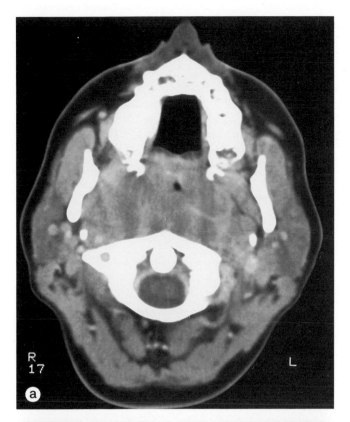

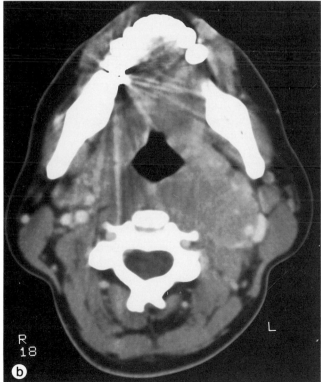

Fig. 6.24 Nasopharyngeal lymphoma. Transaxial postcontrast CT scans demonstrating a huge naso- and oropharyngeal mass. On CT it is difficult to separate pharyngeal disease from associated retropharyngeal lymphadenopathy.

THE OROPHARYNX AND ORAL CAVITY

Imaging modalities

Radiographs

Radiographs are helpful in assessing the overall extent of bone involvement by tumour or infection. An orthopantomogram usually suffices to assess the alveolar ridges and mandible. Additional mandibular views are complementary to the orthopantomogram. Occlusal views of the floor of the mouth provide a view of the soft tissues without bone superimposition and allow detection of calcification in calculi or phleboliths.

A lateral soft tissue radiograph is of limited value in assessing the posterior third of the tongue but is a quick, efficient way of assessing acute epiglottitis.

CT and MRI

At present CT is the preferred imaging modality for the initial assessment of oropharyngeal and oral cavity lesions. It is less sensitive to motion artefacts than MRI, provides a better assessment of adjacent bone involvement and is more widely available. As indicated above there is evidence that CT is better than MRI for nodal staging. However, streak artefacts from dental amalgam and cortical bone margins may cause image degradation on CT.

Indications for CT and MRI examinations

The main indications are:

1. staging of malignant tumours: this is most valuable in lesions more than 2 cm in diameter or those with evidence of deep infiltration – mucosal lesions less than 2 cm in diameter with no palpable nodes usually do not require imaging;
2. the detection of occult primary tumours in patients with unexplained cervical lymphadenopathy or ear pain – approximately 20% of patients with a single palpable malignant lymph node in the mid- or upper cervical chains will have a lesion of the tonsillar region or posterior third of the tongue;
3. the detection of recurrent tumour;
4. assessment of the relationship of benign lesions to adjacent structures prior to surgery.

CT technique

In the majority of cases transaxial images will suffice. Patients are scanned supine, with the neck slightly extended and the head immobilized in an appropriate support. Symmetry of positioning is important. The patients are asked not to cough, swallow or move during scanning and if necessary can hold the tip of the tongue between the teeth. A lateral scout view is obtained and initial 1 cm thick contiguous sections taken from a level just above the hard palate to the hyoid bone during quiet respiration. The plane of section is parallel to the hard palate for lesions on the roof of the mouth and parallel to the horizontal ramus of the mandible for lesions of the tongue or floor of the mouth. The anatomy of the oropharynx is best appreciated in the latter plane of section. Slight angulation may be required to avoid artefacts from dental amalgam.

The unenhanced scan should be assessed for the superior and inferior extent of disease and to determine whether any necessary adjustment to tube angulation should be made. If there is excessive motion artefact then further scans should be obtained during suspended respiration. Enhanced 3–5 mm contiguous sections are then taken through the disease site with 1 cm contiguous sections below this level to the sternal notch. This is important because of the spread of tumours, for example carcinoma of the tongue, to the jugulo-omohyoid node of the inferior deep cervical chain.

Examinations performed to identify occult malignancy should include the nasopharynx and larynx; those for benign lesions can be more limited in extent.

Direct coronal imaging may provide additional, useful information about lesions of the hard palate, peritonsillar region and floor of the mouth. Also, it may be required to avoid artefacts from dental amalgam. Coronal images provide a better anatomical definition of the mylohyoid sling and therefore delineation of the sublingual and submandibular spaces.

All images should be interrogated on the diagnostic console and if necessary image reconstructions should be made from raw data in a smaller field of view and with different algorithms.

MRI technique

While images of the nasopharynx are routinely obtained in the head coil, images of the oral cavity and oropharynx usually require a special neck coil or surface coils. As with the nasopharynx the plane of imaging and pulse sequences chosen will depend on the clinical situation and equipment available. In general, initial multisection T1-weighted images should be obtained to provide an anatomical baseline. For the most part these are obtained in the transaxial plane, although the sagittal plane may be more appropriate for lesions of the posterior third of the tongue. The transaxial plane of section should be similar to that chosen for CT with a slice thickness in the order of 5 mm. Multisection T2-weighted images should then be obtained in an appropriate plane. For lesions involving the palate, floor of mouth, tonsillar fossa and retromolar trigone coronal images are most appropriate.

The use of CT and MR contrast agents is discussed above.

Benign lesions of the oropharynx and oral cavity

Inflammatory lesions – infections

The oropharynx and oral cavity are common sites of infection. Tonsillitis usually involves the palatine tonsils but occasionally the lingual tonsil may be affected in isolation. A peritonsillar abscess, or quinsy, may occur as a complication of tonsillitis. In the oral cavity, gingivitis is frequently encountered.

Radiology has no role in the diagnosis of these conditions, which are readily apparent on visual inspection, but it is frequently of use in the diagnosis and management of retropharyngeal and parapharyngeal abscesses.

Parapharyngeal abscess

Infection of the parapharyngeal space may occur due to direct spread of a peritonsillar abscess, or penetrating trauma. Involvement of the parapharyngeal space often results in widespread dissemination of infection from the skull base down to the submandibular space.

Retropharyngeal abscess

A retropharyngeal abscess occurs most frequently following an upper respiratory tract infection in childhood. Other causes include penetrating injury and tuberculosis, which is thought to spread via the anterior longitudinal ligament of the spine. A retropharyngeal abscess may also occur as a complication of neck surgery (Fig. 6.25).

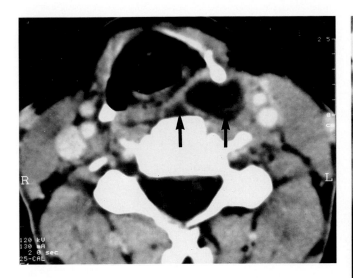

Fig. 6.25 Retropharyngeal abscess. Transaxial contrast-enhanced CT scan demonstrating a multiloculated retropharyngeal abscess (arrows) in a patient following thyroidectomy.

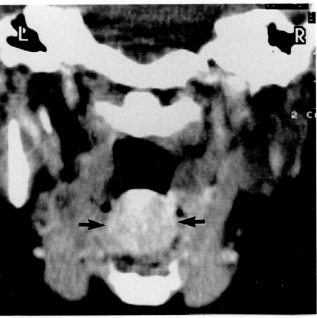

Fig. 6.26 Lingual thyroid. Contrast-enhanced coronal CT scan demonstrating a well-defined midline enhancing mass of lingual thyroid tissue (arrows). (Reproduced by courtesy of Dr JE Gillespie.)

Imaging findings

Severe infection results in the formation of an abscess, which is a thick-walled cavity containing purulent fluid and occasionally gas. Some enhancement of the wall is observed after intravenous contrast. Alternatively, infection may cause cellulitis, leading to diffuse obliteration of tissue planes but no abscess cavity. In both conditions, soft tissue oedema may also occur and is identified as stranding in fat, widening of fat planes and muscle swelling.

Benign masses of the floor of the mouth

Lingual thyroid

A lingual thyroid is undescended thyroid tissue which presents as a midline mass located in the tongue base with variable inferior extension into the postero-medial aspect of the sublingual space. The mass is of high density on CT and shows enhancement after intravenous contrast (Fig. 6.26). On MRI it is of high signal intensity on T1-weighted images, increases in signal intensity on T2-weighted images and will demonstrate enhancement after intravenous contrast. Since up to 75% of patients with a lingual thyroid will have no other functional thyroid tissue [22], a radionuclide thyroid scan is required to confirm the presence of a lingual thyroid and to evaluate the patient for any other thyroid tissue situated within the lower neck.

Haemangioma

Haemangiomas occur more commonly in the tongue base than in the anterior two-thirds of the tongue. Phleboliths may be present on plain films. On CT, haemangiomas have the same attenuation characteristics as muscle but most demonstrate enhancement with contrast. On MRI a haemangioma is of low signal intensity on T1-weighted images and high signal intensity on T2-weighted images. Gradient-recalled echo images may be of value in identifying flow within a haemangioma. Intravenous MRI contrast administration usually causes haemangiomas to enhance.

Lymphangioma

Lymphangiomas [22] are categorized into capillary, cavernous or cystic hygromas according to the size of the cystic spaces within them. Two-thirds of lymphangiomas are seen at birth and over 90% have been diagnosed by 3 years of age. They are found in the neck and upper mediastinum in children but may be located in the cheek, submandibular or sublingual spaces in adults. On both CT and MR they are typically poorly circumscribed, fluid-containing, multiloculated lesions which are laterally located and may infiltrate adjacent structures. Lymphangiomas do not demonstrate enhancement after contrast administration (Fig. 6.27).

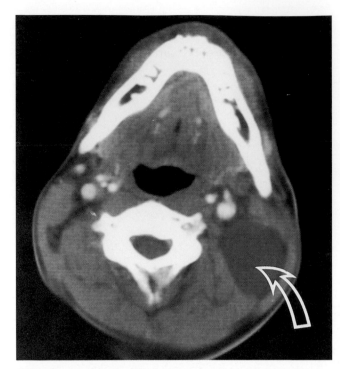

Fig. 6.27 Lymphangioma. Transaxial postcontrast CT scan demonstrating a non-enhancing fluid-density hygroma (open white arrow) situated posterior to the left sternocleidomastoid muscle and carotid sheath. This lesion is unusual in that it appears unilocular.

Dermoids/epidermoids

These [23] most commonly present as a midline neck mass anywhere between the submental region and suprasternal notch. Twenty per cent occur in the mouth, either in the sublingual or submental spaces. They are commonly unilocular, fluid- or fat-containing lesions on CT or MRI. A chemical shift artefact may be observed on MRI images when fat and fluid elements are contained within a dermoid.

Ranula

A ranula is a unilocular cyst which arises in the floor of the mouth and is believed to represent a mucus retention cyst of the sublingual gland. Ranulas may be simple, i.e. confined to the sublingual space, or plunging, when they extend through the mylohyoid muscle into the submandibular space to form a pseudocyst. On CT they will be of fluid attenuation, and on MRI their proteinaceous content will result in very high signal intensity on T2-weighted images.

Benign tumours of the floor of the mouth

Rare benign tumours occurring in the floor of the mouth include lipoma, pleomorphic adenoma, rhabdomyoma, schwannoma and neurofibroma.

Second branchial cleft cyst

This type of branchial cyst [24] is slightly more common in males and accounts for up to 95% of all branchial cleft cysts. It presents as a mass at the angle of the mandible, from which a fistula or sinus tract may drain into the anterior neck above the clavicle. On CT and MRI branchial cleft cysts are typically unilocular, well-circumscribed and non-septated lesions, situated postero-inferior to the submandibular glands and antero-medial to the sternocleidomastoid muscles (Fig. 6.28). The carotid sheath lies posteriorly and part of the branchial cyst may protrude between the internal and external carotid arteries.

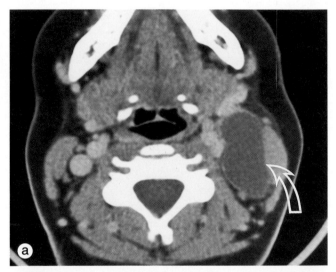

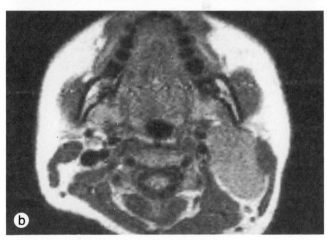

Fig. 6.28 Branchial cleft cyst. (a) Transaxial postcontrast CT image. **(b)** Transaxial T1-weighted MR image. The unilocular branchial cleft cyst (open arrow) is located posterior to the left submandibular gland, medial to the left sternocleidomastoid muscle and lateral to the carotid sheath. The high signal intensity of the cyst on the MR image may be due to very high protein content or haemorrhage.

Suprahyoid thyroglossal cyst [23]

Only about 25% of thyroglossal cysts occur in the suprahyoid region and a tiny percentage occur in the posterior third of the tongue. They present as midline masses which move on swallowing and often cause fullness in the floor of the mouth. Half are diagnosed in children under 10 years of age. On CT and MRI they are well-circumscribed, midline or paramedian lesions situated between the anterior bellies of the digastric muscles and they demonstrate rim enhancement after intravenous contrast (Fig. 6.29).

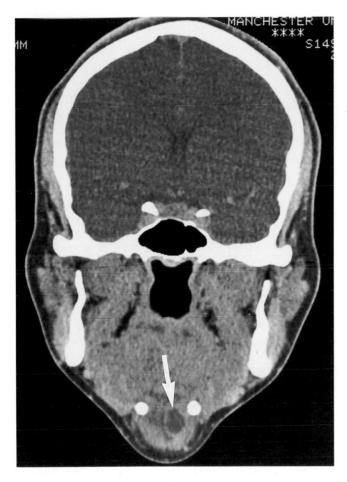

Fig. 6.29 Thyroglossal cyst. Coronal enhanced CT image showing a paramedian thyroglossal cyst (arrow) with rim enhancement.

Malignant tumours of the oropharynx and oral cavity

The aetiology, pathology, role of imaging and treatment of pharyngeal malignancies in general has been discussed.

Clinical features

Patients typically present with a sore throat, pain on swallowing, referred earache or trismus. Oral cavity tumours are often identified on visual inspection. More advanced lesions may cause neurological symptoms by perineural infiltration or involvement of cranial nerves IX to XII in the parapharyngeal space. Some patients have an asymptomatic primary lesion but develop a metastatic lymph node mass within the neck.

Tumour staging is according to the TNM classification (Tables 6.4 and 6.5). Nodal and distant metastases are staged according to the TNM classification in Table 6.3.

Routes of tumour spread

Tumours spread superficially within the mucosa and this is identifiable on inspection. Deep spread occurs along and into adjacent muscles and spaces, such as the parapharyngeal or sublingual space. Tumours may also infiltrate the periosteum of bone and penetrate into the medullary cavity via neurovascular foramina. Perineural infiltration may occur within the soft tissues. To a large degree, the site of the primary lesion governs its route of spread into adjacent structures and this will be discussed in more detail with reference to specific primary tumour sites.

Table 6.4   TNM staging classification: oropharynx

***T* – Primary tumour**

TX	Primary tumour cannot be assessed
T0	No evidence of primary tumour
Tis	Carcinoma *in situ*

Oropharynx

T1	Tumour 2 cm or less in greatest dimension
T2	Tumour more than 2 cm but not more than 4 cm in greatest dimension
T3	Tumour more than 4 cm in greatest dimension
T4	Tumour invades adjacent structures, e.g. through cortical bone, soft tissues of neck, deep (extrinsic) muscle of tongue

---

Table 6.5   TNM staging classification: oral cavity

T – Primary tumour

TX	Primary tumour cannot be assessed
T0	No evidence of primary tumour
Tis	Carcinoma *in situ*
T1	Tumour 2 cm or less in greatest dimension
T2	Tumour more than 2 cm but not more than 4 cm in greatest dimension
T3	Tumour more than 4 cm in greatest dimension
T4	Tumour invades adjacent structures, e.g. through cortical bone, into deep (extrinsic) muscle of tongue, maxillary sinus, skin

---

CT and MR appearances

In oropharyngeal carcinoma CT or MRI findings [6,25] may include:

1. an exophytic or ulcerating mass;
2. asymmetry or thickening of the tonsillar pillars;
3. reduction in the oropharyngeal lumen due to circumferential tumour infiltration;
4. palatal perforation;
5. infiltration of adjacent soft tissues and loss of fascial planes;
6. tumour extension into the parapharyngeal, retropharyngeal or masticator spaces;
7. superior extension into the nasopharynx, inferior spread to involve the base of tongue or anterior infiltration of the oral cavity;
8. Lymphadenopathy.

CT and MRI findings in oral cavity carcinoma may include:

1. the identification of a mass together with its local spread;
2. involvement of the gingivobuccal sulci;
3. inferior extension into the floor of mouth, and the associated sublingual and submandibular spaces;
4. superior extension of disease through the hard palate into the nasal cavity or maxillary antra;
5. posterior extension into the oropharynx;
6. obstruction of the submandibular or parotid glands due to involvement or compression of their ducts.

Tumours of the oropharynx

One third of all pharyngeal carcinomas occur in the oropharynx. The distribution of disease is as follows: tonsils/tonsillar pillars 50%, base of tongue 20%, soft palate 10%, valleculae and lingual epiglottis 10%, posterior wall 5% and lateral wall 5%.

Specific features of oropharyngeal tumours related to their site

1. Anterior tonsillar pillar: These lesions have a good prognosis because they are often diagnosed early. Primary routes of spread are to the soft palate, the posterior aspect of the hard palate and the maxillary antra. Antero-laterally disease extends to the retromolar trigone and thence to the buccal mucosa. Large lesions may involve the tongue.

2. Tonsillar fossa: Tumours may arise within the tonsil itself or in the overlying mucous membrane. Early spread occurs to the posterior tonsillar pillar and the oropharyngeal wall and there is involvement of the glossotonsillar sulcus and the base of the tongue in a quarter of patients (Fig. 6.30). Spread occurs through the superior constrictor muscle of the lateral oropharynx into the parapharyngeal space and from there to the base of the skull. Inferior spread can occur below the pharyngoepiglottic fold into the pyriform fossa (Fig. 6.31). Because of the

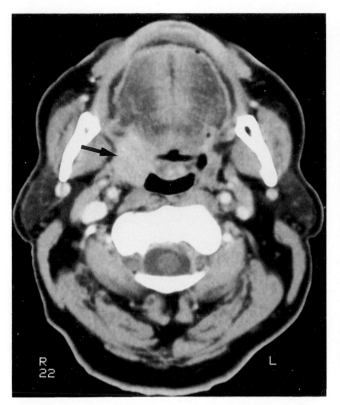

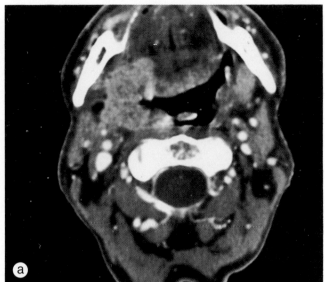

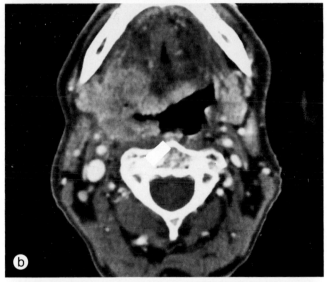

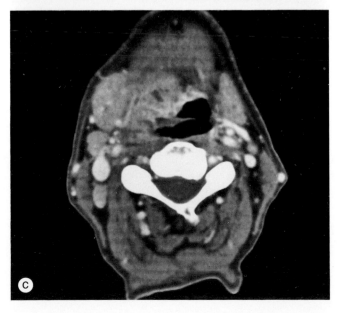

Fig. 6.30   Tonsillar fossa carcinoma involving the base of the tongue. Enhanced transaxial CT scan through the tonsillar fossa demonstrating a right tonsillar fossa mass (arrow) with early invasion of the base of the tongue.

Fig. 6.31   Extensive tonsillar fossa carcinoma. Transaxial postcontrast CT images from the angle of the mandible to the valleculae. The right-sided tumour involves the oropharyngeal wall, the base of the tongue and the parapharyngeal space. There is infiltration of the superior aspect of the right submandibular gland and disease extends into the valleculae. Significant right-sided lymphadenopathy is present (arrow).

early local extension of disease, tonsillar lesions are not amenable to local resection and radiotherapy is employed. Salvage surgery can be done in the case of recurrence but this is extensive, involving mandibulectomy, resection of the tonsil and both anterior and posterior pillars and also part of the tongue.

3. Posterior tonsillar pillar: Tumours in this region are uncommon and spread occurs inferiorly along the palatopharyngeal muscle.

4. Base of the tongue: Tumours of the base of the tongue characteristically show early deep infiltration of adjacent structures and this accounts for their low cure rate of only 15%. Involvement of the valleculae eventually leads to penetration through the hyoepiglottic ligament into the pre-epiglottic space. More inferior spread may occur to the larynx. Tumours involving the lateral base of the tongue extend into the glossotonsillar sulcus and from there into the neck (Fig. 6.32). Anterior extension will involve the oral tongue. Treatment is by local resection or radiotherapy, but more extensive tumours require hemiglossectomy. Since surgery requires preservation of one lingual artery and hypoglossal nerve, cross-sectional imaging, particularly CT, is useful in identifying whether these structures have been spared.

5. Soft palate: Tumours occur on the oral surface of the soft palate and on inspection there may be multiple sites of disease with normal appearing mucosa between. Spread occurs to the tonsillar pillars, hard palate, superior constrictor muscle, medial pterygoid muscle and eventually to the base of the skull. Advanced tumours may cause perforation of the palate or extend superiorly along the lateral walls of the nasopharynx.

Tumours of the oral cavity

The floor of the mouth

Ninety per cent of lesions occur within 2 cm of the anterior midline and often demonstrate early penetration into the adjacent genioglossus and geniohyoid muscles. Extension to the gingiva and periosteum of the mandible is another early feature whereas actual invasion of the mandibular cortex is a late phenomenon (Fig. 6.33). Sometimes floor of mouth tumours obstruct the submandibular duct.

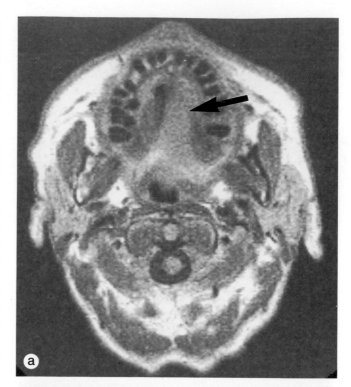

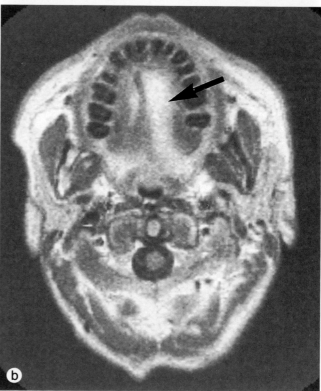

Fig. 6.32 Anterior extension of base of tongue carcinoma. Transaxial T1-weighted MR images pre- and postcontrast gadolinium DTPA. There is a relatively abnormal high signal intensity infiltrative tumour in the base of the tongue (arrow) which extends anteriorly predominantly on the left side. There is marked tumour enhancement after gadolinium DTPA administration.

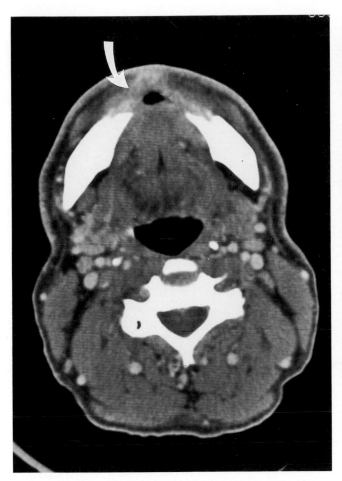

Fig. 6.33 Carcinoma of the floor of the mouth.
Transaxial postcontrast CT image demonstrating an ulcerated anterior tumour (arrow) which is eroding the mandible. Small bilateral deep cervical lymph nodes are present.

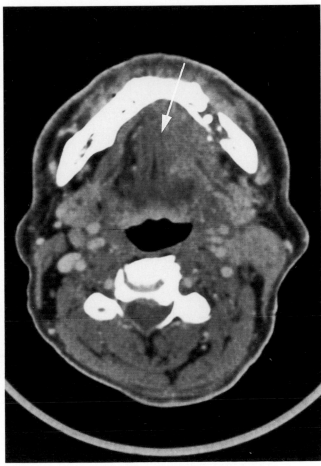

Fig. 6.34 Carcinoma of the tongue. Transaxial postcontrast CT scan demonstrating a mass in the left tongue which is infiltrating the left genioglossus muscle (arrow) and displacing the genioglossus and lingual septum to the right. There is also tumour extension into the mandible.

The anterior two-thirds of the tongue

A tumour of the anterior third is the least common site of tongue malignancy but 80% are cured due to early presentation. Initial spread is into the floor of the mouth. Middle third lesions invade the tongue musculature and then the lateral floor of the oral cavity (Fig. 6.34). In these patients sagittal or coronal scans are useful to determine the extent of any deeply infiltrating tumour and the presence of invasion of the base of tongue so as to facilitate adequate tumour resection.

The buccal mucosa

Tumours of the buccal mucosa are rare and are most frequently low-grade squamous cell carcinomas. They may obstruct the parotid duct or infiltrate the buccinator and masseter muscles, causing trismus.

Gingiva/hard palate/retromolar trigone tumours

Lower gum tumours infiltrate the buccal mucosa and floor of mouth with early involvement of the periosteum of the mandible.

Hard palate tumours are most commonly adenoid cystic carcinomas and may infiltrate through the hard palate into the nasal cavity or maxillary antra (Fig. 6.35). Lateral extension may occur into the gingivobuccal sulcus (Fig. 6.36) and posterior extension into the oropharynx (Fig. 6.37).

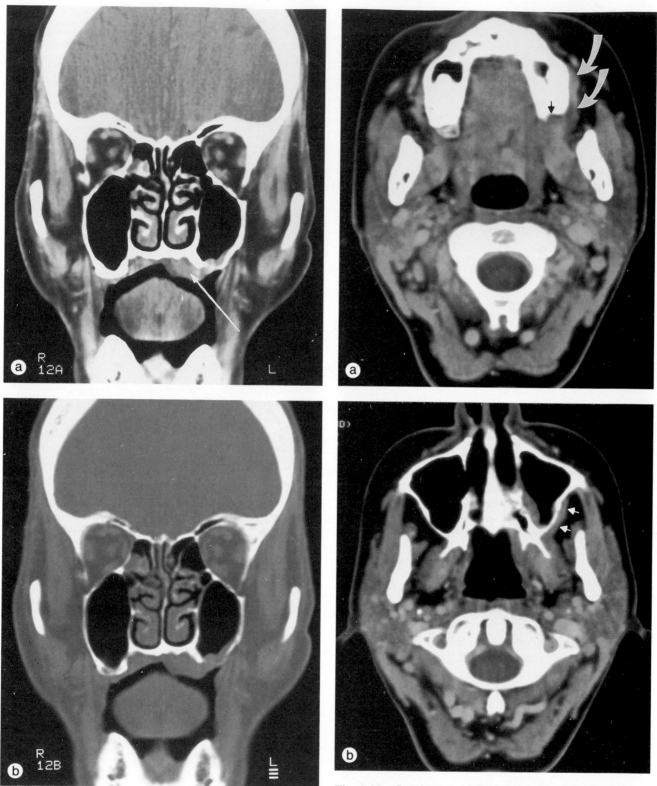

Fig. 6.35 Carcinoma of the hard palate extending into the left maxillary antrum. Coronal postcontrast CT scans on (**a**) soft tissue and (**b**) bone settings. The small left hard palate adenocystic carcinoma extends to the midline (arrow). There is early erosion of the floor of the left maxillary antrum.

Fig. 6.36 Carcinoma of the hard palate involving the gingivobuccal sulcus. Transaxial contrast-enhanced CT scans demonstrating posterior erosion of the left superior alveolar ridge (black arrow) and anterior tumour extension into the gingivobuccal sulcus (curved arrows). There is also tumour extension around the lateral wall of the left maxillary antrum (small arrows on (**b**)).

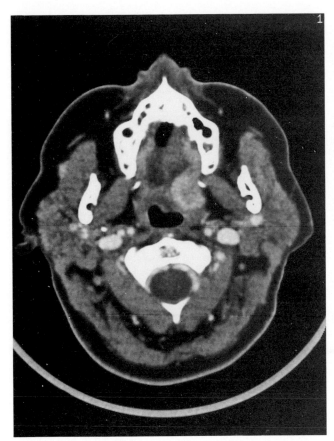

Fig. 6.37 Carcinoma of the hard palate with posterior extension. Transaxial postcontrast CT scan showing enhancing carcinoma of the hard palate extending posteriorly into the soft palate and postero-laterally into the left oropharynx and upper tonsillar fossa.

Retromolar trigone lesions invade the buccal mucosa, the anterior tonsillar pillar and the mandible, although bone destruction is uncommon. Posterior extension can occur to the parapharyngeal space with involvement of the medial pterygoid muscles (Fig. 6.38). Postero-lateral extension is through the buccinator muscle. Superior extension involves the maxilla lymphoma.

The oropharyngeal components of Waldeyers ring may be involved by lymphoma. This has been discussed on page 91.

TUMOUR RECURRENCE

Recurrent squamous cell carcinoma usually presents within 2 years of the primary lesion, although non-keratinizing tumours can recur several years after treatment (Figs 6.39 and 6.40). Local recurrence is more likely to occur at the resection margins in patients who have undergone surgery and in the

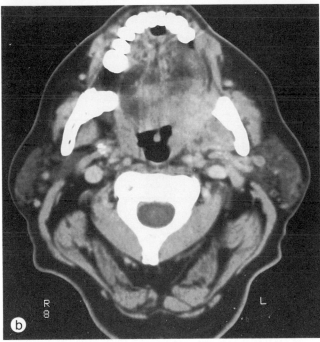

Fig. 6.38 Retromolar trigone carcinoma. Transaxial postcontrast CT scans demonstrating a left retromolar tumour extending into the tonsillar fossa and laterally into the left medial pterygoid muscle.

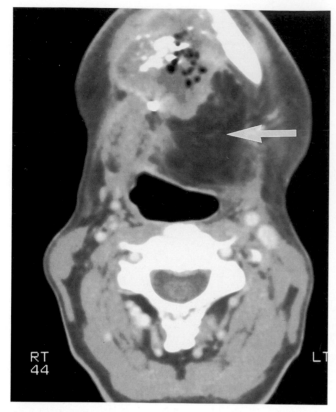

Fig. 6.39 Recurrent squamous cell carcinoma of the tongue after total glossectomy and hemimandibulectomy. Transaxial postcontrast CT scan showing a large necrotic anterior recurrent tumour. There has been a flap reconstruction of the floor of the mouth which is shown as a fatty structure (arrow).

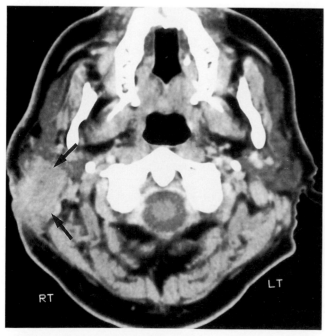

Fig. 6.40 Recurrent squamous cell carcinoma of pinna. Transaxial postcontrast CT scan showing a relatively well-defined right-sided recurrent tumour mass (arrows).

centre of a lesion treated by radiotherapy. Patients typically develop pain, and trismus may occur secondary to involvement of the infratemporal fossa muscles (Fig. 6.41). Lymphadenopathy may be ipsi-lateral or contralateral. The latter occurs more frequently in patients who have undergone surgery or

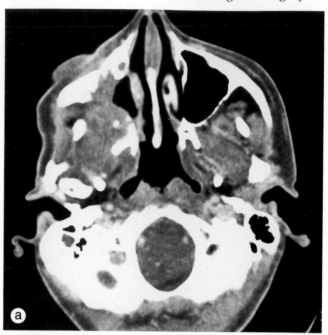

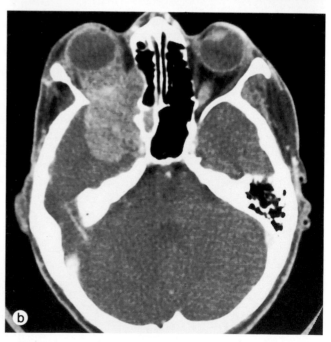

Fig. 6.41 Recurrent adenocystic carcinoma causing trismus. Transaxial postcontrast CT scans (**a**) through the infratemporal fossa and (**b**) through the middle cranial fossa. An enhancing tumour recurrence is seen involving the right maxilla, cheek, infratemporal fossa musculature and temporomandibular joint. There is extension into the right orbit and middle cranial fossa.

radiotherapy, when the submental nodes provide a route for contralateral spread. Clinical examination is often difficult due to scarring from surgery, radiotherapy or a combination of the two.

Although there are recognized difficulties in distinguishing between recurrent tumour, fibrosis and inflammation, CT and MRI have a role in suspected recurrent disease, particularly when further treatment such as salvage surgery or palliative radiotherapy is contemplated. After surgery, the resection margins are typically well defined and the adjacent normal anatomy shows no distortion [26]. Extensive surgical reconstruction may have been undertaken (Fig. 6.39 and 6.42). Surgical flaps may have been

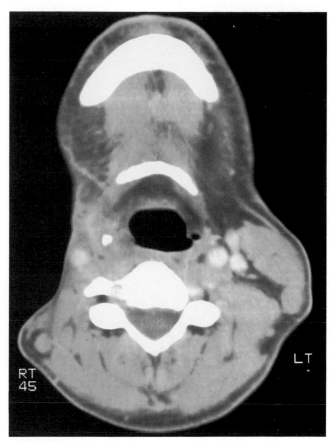

Fig. 6.43 Appearances following radical neck dissection and radiotherapy for squamous cell carcinoma of the tongue. The right sternocleidomastoid muscle, internal jugular vein, lymph nodes and submandibular gland have been resected. There is thickening of the skin, stranding of the subcutaneous fat and amorphous soft tissue around the right carotid artery. Further characterization of the lesion is impossible and it requires needle biopsy.

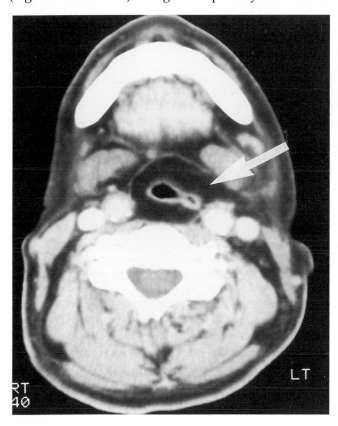

Fig. 6.42 Neopharynx. Transaxial postcontrast CT scan shows a neopharynx with fatty atrophy of the muscular component (arrow).

inserted and these are typically fatty, sometimes with an area of apparent soft tissue thickening, corresponding to the vascular pedicle. Occasionally scarring does cause obliteration of surgical planes, particularly when there has been superadded infection.

After radiotherapy, the picture may be more confusing due to oedema of the residual normal tissues, marked stranding within fat spaces and thickening of the skin and subcutaneous tissues (Fig. 6.43). Changes may occur in any tissue within the

treatment field (Fig. 6.44). It may be necessary to have sequential examinations to evaluate patients for early recurrence or to perform percutaneous biopsy in those patients where the results of imaging remain equivocal. The presence of an obvious mass lesion (Fig. 6.39 and 6.40) or new lymphadenopathy is almost invariably due to tumour.

MRI does help to differentiate tumour from fibrosis. Fibrotic tissue is typically of low signal intensity on both T1- and T2-weighted images whereas tumour exhibits high signal intensity on T2-weighted images [27]. However, it has been shown that infiltrating tumours which excite a fibrotic response may be misdiagnosed as fibrosis on MRI and, equally, hypervascular or inflammatory lesions may be wrongly assumed to represent recurrent tumour.

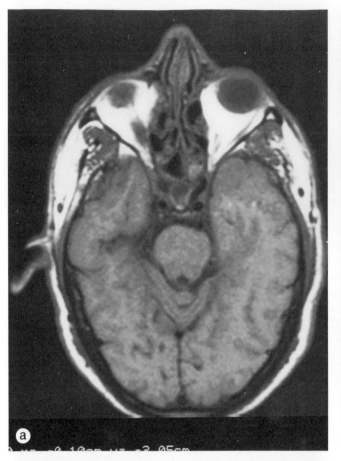

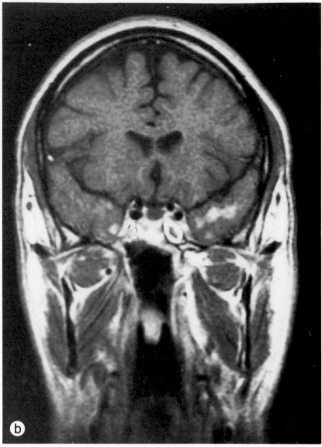

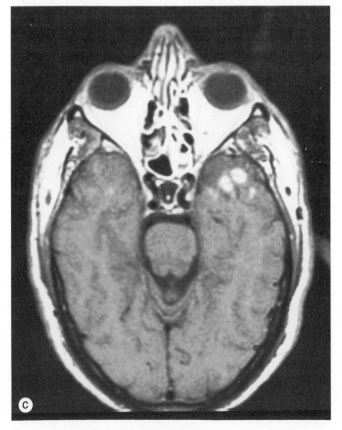

Fig. 6.44 Radiation change within both temporal lobes after treatment for an extensive nasopharyngeal carcinoma. (**a**) T1-weighted precontrast transaxial image. (**b**) T1-weighted postcontrast transaxial image. (**c**) T1-weighted postcontrast coronal image. Focal areas of enhancement are seen in both temporal lobes, predominantly involving the white matter. The treatment field encompassed both temporal lobes. The patient is asymptomatic.

THE SALIVARY GLANDS

Anatomy [1,2,28]

Parotid gland

The parotid glands are irregular structures whose shape most resembles an inverted pyramid. They are situated below the external auditory meatus between the mandible and sternocleidomastoid muscle. The concave superior surface of each gland is bordered by the external auditory meatus and the posterior aspect of the temporomandibular joint. The inferior apex overlies the posterior belly of the digastric muscle and the base of the carotid triangle. Antero-medially the surface of each gland is grooved by the posterior border of the mandibular ramus and covers the postero-inferior surface of the masseter muscle, lateral temporomandibular joint and mandible. The postero-medial surface of the parotid gland is moulded to the mastoid process, sternocleidomastoid muscle, digastric muscle (posterior belly) and styloid process. The lateral superficial margin of the parotid is covered by skin and fascia and also has the superficial parotid lymph nodes embedded in its surface.

Several important structures pass through the parotid gland. The external carotid artery enters into the postero-medial gland and divides into the maxillary artery and superficial temporal artery. The retromandibular vein, formed by the junction of the maxillary and superficial temporal veins, is also found on the postero-medial surface of the parotid gland and joins the posterior auricular vein to form the external jugular vein. The auriculotemporal nerve is embedded in the capsule of the parotid's superior surface.

On transaxial sections, a line drawn from the stylomastoid foramen which passes immediately lateral to the retromandibular vein will divide the parotid into superficial and deep portions. The facial nerve lies in the most superficial portion of the gland. It leaves the skull via the stylomastoid foramen and enters the parotid immediately lateral to the styloid process, anterior to the posterior belly of the digastric muscle and medial to the tip of the mastoid process. The facial nerve then passes down behind the mandibular ramus. Within the parotid gland, the facial nerve lies lateral to the retromandibular vein, displacement of which is readily seen on unenhanced and enhanced CT scans. A line drawn between the posterior belly of the digastric muscle and the sternocleidomastoid muscle to the lateral border of the mandible approximates the position of the facial nerve in the superior aspect of the gland. At lower levels the nerve will lie 8–9 mm behind the posterior aspect of the mandibular ramus.

The parotid duct passes from the anterior aspect of the gland across the masseter muscle, where a small amount of accessory parotid tissue drains into it via the accessory parotid duct, and then turns medially at the anterior border of the masseter (Fig. 6.45). The duct passes obliquely forwards between the buccinator muscle and oral mucosa to open into the oropharynx at the level of the upper second molar tooth.

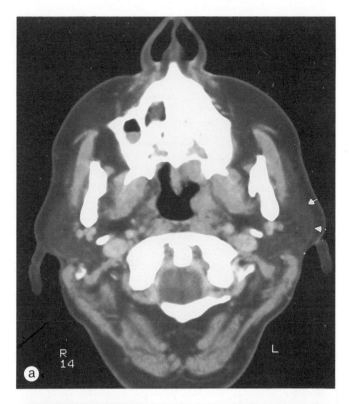

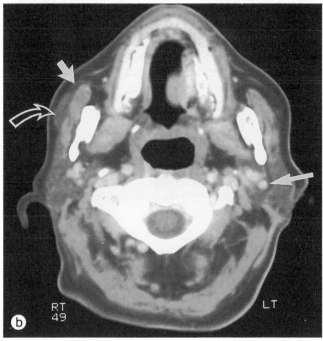

Fig. 6.45 The normal parotid gland and duct. Transaxial postcontrast CT scans showing: (1) the relatively low-density parotid gland (small arrows in (**a**)); (2) the retromandibular vein (long arrow in (**b**)); (3) accessory parotid tissue (open arrow in (**b**)); (4) the parotid duct (short arrow in (**b**)).

Vascular supply

The parotid gland is supplied by the external carotid artery and its draining veins empty into the external jugular vein.

Lymph nodes

In and around the parotid gland are superficial and deep lymph nodes, which receive lymphatic drainage from the parotid. The superficial group of lymph vessels and nodes also drain the frontotemporal region of the scalp, the eyelids and the anterior auricle with efferent drainage to superficial cervical nodes. The deep group lies within the substance of the gland and also drains the eustachian tube, external auditory canal and deep tissues of the face. The efferent channels are to subparotid and deep cervical nodes.

Submandibular gland

The submandibular gland has a large superficial portion located in the submandibular space and a small deep portion situated within the posterior sublingual space. These join posteriorly along the free edge of the mylohyoid muscle. The submandibular duct passes from the antero-medial border of the gland over the posterior surface of the mylohyoid muscle to reach the frenulum of the tongue. The lingual artery indents the submandibular gland laterally and multiple lymph nodes are situated on the surface of the gland.

Vascular supply

The submandibular glands are supplied by branches of the facial and lingual arteries, and venous drainage is via the facial and lingual veins.

Lymph nodes

The submandibular gland does not contain any intraglandular nodes but there are a number of nodes situated anterior to the gland, which if enlarged can appear intraglandular. The submandibular lymph nodes receive drainage from the submandibular and sublingual glands, the submental nodes, the anterior two-thirds of the tongue, the upper lip and lateral segments of the lower lip and the external nose.

Sublingual glands

Each sublingual gland is situated on the lingual aspect of the mandible close to the symphysis menti within the sublingual space. The oral mucosa is located immediately above the gland, the mylohyoid muscles below and the deep portion of the submandibular gland posteriorly, with the mandible laterally and the genioglossus muscle medially. Multiple execretory ducts drain each sublingual gland.

Vascular supply

The sublingual and submental arteries supply the sublingual glands and venous drainage is via the sublingual and submental veins.

Imaging modalities

While sialography may reveal the presence of a mass by displacement or non-filling of part of the ductal system, cross-sectional imaging techniques enable mass lesions to be better assessed. A distinction can be made between intra- and extraglandular tumours, lesions can be better characterized, the relationship of parotid tumours to the facial nerve defined, involvement of adjacent structures assessed and metastatic lymphadenopathy identified.

CT technique

The parotid gland

An initial scout view should be obtained followed by unenhanced 5 mm contiguous sections in the transaxial plane from the superior aspect of the gland to the hyoid bone. The plane of section may need to be altered if artefacts from dental amalgam are predicted, in which case an appropriate semi-axial or coronal plane can be defined on the scout view. Ideally the patient should be asked not to swallow or breathe during data acquisition.

The submandibular gland

The basic protocol is as above with sections covering the sublingual and submandibular spaces.

Contrast-enhanced CT scans

Contrast enhancement is often unnecessary, particularly in the assessment of the parotid gland. This gland is of low density relative to adjacent soft tissues, the submandibular gland being of similar density to adjacent soft tissue. However enhanced scans can be of value:

1. to distinguish between lesions of the deep lobe of the parotid and parapharyngeal space;
2. to assess vascular and nodal structures adjacent to the salivary gland;
3. to assess a dense parotid gland.

CT sialography

The use of intraductal contrast medium combined with CT may be of value in the following situations:

1. **Parotid gland** – when there is a strong clinical suspicion of disease but conventional CT scanning has proved negative or equivocal;
2. **Submandibular gland** – in the evaluation of possible mass lesions.

CT-guided biopsy

CT enables adequate localization for percutaneous needle aspiration biopsy if required.

Magnetic resonance imaging

At present CT is the mainstay for evaluation of glandular and periglandular masses but MRI [28,29] may prove to be more valuable, particularly in the parotid region. It can readily distinguish between lesions in the deep lobe of the parotid and in the parapharyngeal space, but more significantly, the facial nerve is often seen distal to the stylomastoid foramen thereby allowing a more accurate assessment of its relationship to mass lesions.

Identification of the fat plane between the gland and an extrinsic mass lesion will usually require short TR/TE images with the parotid gland being hypointense relative to the fat plane. Although the facial nerve can be identified on head coil images, the use of surface coils will provide greater detail. On short TR/TE images the nerve and its branches are hypointense relative to the parenchyma (Fig. 6.46).

A disadvantage of MRI is the relatively poor detection of intraglandular calcification or of calculi within the salivary ducts. CT is therefore preferred to MRI in cases of suspected infection.

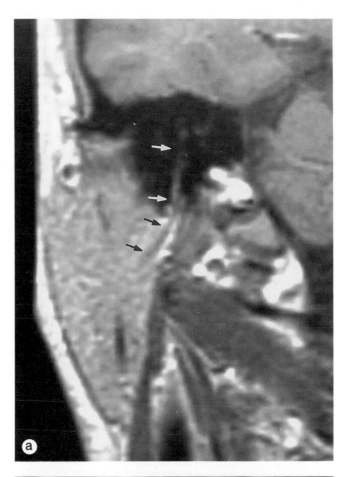

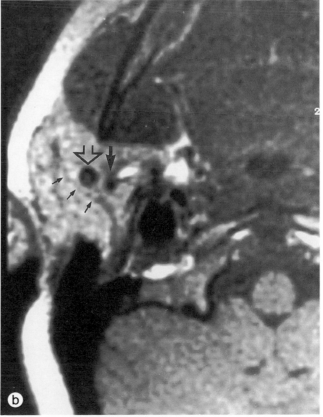

Fig. 6.46 Position of the facial nerve in the parotid gland. (a) Coronal-oblique T1-weighted image (following gadolinium DTPA, given for other purposes). The third part of the facial nerve is seen descending through the signal void of the petrous bone (white arrows), exiting the stylomastoid foramen and entering the parotid gland (black arrows). **(b)** Axial-oblique T1-weighted image. The characteristic S-shaped curve of the seventh nerve (small black arrows) as it courses anterolaterally behind the retromandibular vein (open arrow) and external carotid artery (large solid arrow) is well demonstrated. Both images were obtained using a standard head coil. (Reproduced by courtesy of Dr JPR Jenkins and Dr JE Gillespie.)

Technique

The choice of coil will depend on the area to be scanned, the need to compare between the normal and abnormal side, the anatomical detail required and coil availability. T1- and T2-weighted images should be obtained in appropriate planes which can be assessed from localizer images. The most commonly used planes for both parotid and submandibular glands will be the transaxial and coronal. Section thickness should be in the order of 5 mm and patients should be asked to reduce swallowing to a minimum.

Image evaluation

Images should be evaluated for:

1. the presence of sialoliths;
2. glandular size and CT/MR characteristics;
3. the relationship of parotid lesions to the facial nerve;
4. whether mass lesions are intra- or extraglandular;
5. the presence or absence of tissue planes adjacent to the salivary glands;
6. the presence of lymphadenopathy.

Sialoliths

These may be responsible for glandular enlargement or atrophy if there has been longstanding ductal obstruction. Ductal obstruction may lead to proximal inflammatory changes, with subsequent spread beyond the gland.

Glandular size/CT and MRI characteristics

The gland may atrophy due to long-standing obstruction or radiotherapy.

Glandular enlargement may be the result of:

1. obstruction secondary to ductal stricture;
2. inflammatory disease, related either to obstruction or to an ascending bacterial or viral infection;
3. tumours, either benign or malignant, and intraglandular lymphadenopathy;
4. sarcoidosis;
5. Sjögren's syndrome;
6. compensatory hypertrophy after radiotherapy or resection of the contralateral gland.

Although the CT and MRI findings do not allow a tissue-specific diagnosis, features such as border definition, pattern of contrast enhancement and whether the lesion is cystic or solid narrow the differential diagnosis. Generally, round, well-defined lesions are likely to be benign and calcification is found most frequently in benign lesions. When a mass is lobulated or irregular in contour and has heterogeneous density or signal intensity, or central necrosis, it is more likely to be malignant. Other features indicating a malignant lesion are intraglandular duct obstruction without evidence of calculi, cervical lymphadenopathy and bone invasion. On MRI, more aggressive tumours may have low signal intensity on both T1- and T2-weighted images.

Mass lesions and their relationship to the facial nerve

Clinically palpable mass lesions around the parotid gland may arise within the superficial lobe, within the deep lobe, in the parapharyngeal space, or superficial or lateral to the gland.

The distinction between lesions in the deep lobe of the parotid and the parapharyngeal space is based on the identification of a fat plane between the mass and a normal-appearing parotid gland. If this is absent then it may not be possible to distinguish between a lesion arising in the parotid gland which has extended into the parapharyngeal space and *vice versa*.

The surgical approach to lesions within the parotid is dependent on their position, lesions in the superficial lobe being removed by superficial parotidectomy and those in the deep lobe by total parotidectomy. The latter involves significant risk to the facial nerve, the course and relationships of which are described above. Therefore, it is important to assess the relationship of any mass lesion to the stylomastoid foramen, particularly whether there is an adequate margin of normal parotid tissue (approximately 0.5 cm) adjacent to it. If not, then partial mastoidectomy may be required.

Lymph nodes

Focal or diffuse enlargement of the parotid or submandibular gland can be due to peri- or intraglandular nodal enlargement. The nodal enlargement in turn can be due to a systemic disease or as a response to inflammatory or neoplastic processes in the regions they drain.

On unenhanced CT scans these lymph nodes are relatively isodense with respect to parotid tissue. On enhanced scans the nodes appear hypodense with respect to the enhancing parotid parenchyma.

Benign lesions of the salivary glands

Inflammatory

Chronic infection

Chronic infection occurs secondary to duct strictures and resullts in increased density of the gland on CT, sometimes with punctate calcification. Where only a portion of the gland is obstructed, a mucous cyst may form.

Granulomatous disease

Sarcoid is the most common granulomatous condition to affect the salivary glands, and usually involves the parotid. Involved glands show patchy increase in density and there may be fatty replacement of glandular tissue. Tuberculosis tends to affect the salivary glands from contiguous lymph node spread.

Autoimmune disorders

Sjögren's disease is the autoimmune condition which particularly affects the salivary glands. It may be primary, in association with keratoconjunctivitis sicca, or secondary occurring in conjunction with other connective tissue disorders such as rheumatoid arthritis, systemic lupus erythematosus, scleroderma and polymyositis. On CT, Sjögren's disease causes some overall increase in density of the affected glands, within which punctate low-density lesions are identified.

Sialectasis and salivary gland stones

Sialectasis occurs when the alveoli within salivary glands coalesce to form cystic spaces. Debris from the damaged alveoli blocks the salivary ducts, leading to focal areas of stenosis and proximal duct dilatation with intermittent swelling of the gland. Stones form around the epithelial debris and are more common in the submandibular than the parotid gland, due to the higher calcium content of the secretions of the former. Patients present with glandular swelling or pain and may suffer from secondary infection. Radiographs can be used to identify calcification – usually located in the line of the duct – and sialography may demonstrate an obstructed duct or sialectasis with cystic spaces, focal duct stenosis and dilatation.

Benign tumours

Pleomorphic adenomas

These are the commonest tumours, accounting for approximately 75% of benign lesions of the parotid and 50% of benign lesions of the submandibular gland. In the parotid the majority of lesions occur in the superficial lobe and morphologically they are well-circumscribed mass lesions with a fibrous capsule which may be deficient in some areas. On histology these lesions have varying amounts of glandular, ductal and squamous epithelial components within a mesenchymal stroma of varying pattern. The stroma is usually infiltrated with connective tissue mucin and occasionally there is metaplasia to cartilage. The combination of epithelial tissue with cartilage gave rise to the old name of 'mixed parotid tumour'. If incompletely removed, pleomorphic adenomas have a tendency to recur locally. Five to ten per cent of these adenomas contain foci of malignancy, which may be of the epithelial or stromal components.

On CT the majority of lesions are well defined and isodense with normal parotid tissue. They demonstrate enhancement, which is usually homogeneous (Fig. 6.47), although areas of necrosis may be demonstrated. Indicators of malignancy are an indistinct border, although this may be seen with associated inflammation, and lesions with low density centres and a thin enhancing rim.

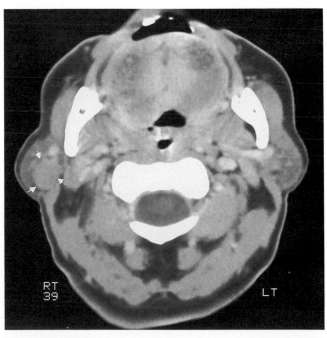

Fig. 6.47 Pleomorphic adenoma. Transaxial postcontrast CT scan demonstrating a well-defined tumour (arrows) in the superficial portion of the right parotid gland.

On MRI normal parotid tissue has an intermediate signal intensity between muscle and fat. On T1-, T2- and proton-density-weighted images the gland is hyperintense relative to muscle and hypointense relative to fat. Benign salivary tumours in general have a relatively high signal intensity on T2-weighted images. The tumour/normal gland interface is usually better seen on T2-weighted or postcontrast images. Pleomorphic adenomas are predominantly homogeneous and smoothly marginated, but they may contain focal areas of calcification and necrosis (Fig. 6.48).

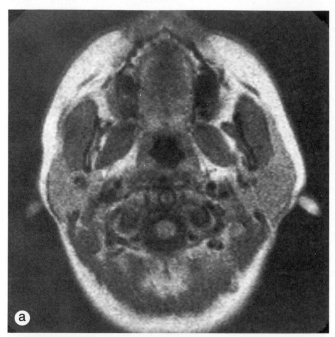

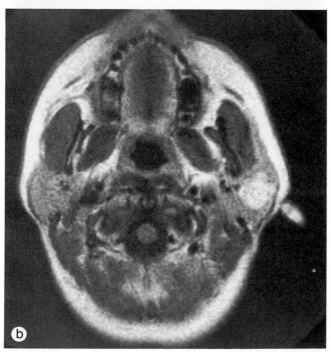

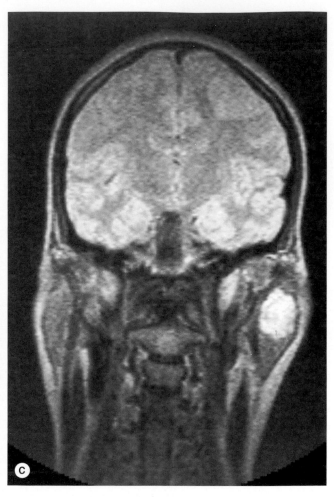

Fig. 6.48 Pleomorphic adenoma. (a) Transaxial T1-weighted MR image. **(b)** Transaxial T1-weighted postcontrast MR image. **(c)** Coronal T2-weighted MR image. It is difficult to identify the right-sided parotid tumour on the precontrast T1-weighted image but it is causing slight enlargement of the gland. On the postcontrast T1-weighted image the tumour enhances markedly and is lobulated but appears better defined.

Warthin's tumour (adenolymphoma)

Warthin's tumour is the second commonest benign tumour involving the salivary glands. It occurs in male patients over 40 years of age, most frequently in the parotid gland. Multiple bilateral tumours may occur. Histologically, this tumour is a papillary cystadenoma with lymphocytes in the supporting stroma.

On CT these tumours are usually hyperdense with respect to the normal parotid parenchyma, the cystic areas being poorly defined (Fig. 6.49).

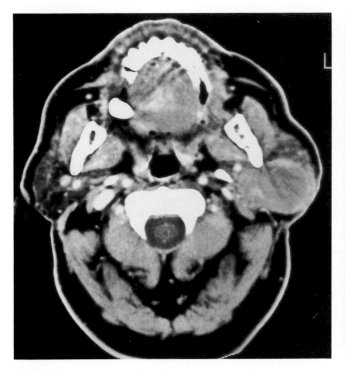

Fig. 6.49 Warthin's tumour. Contrast-enhanced transaxial CT scan showing a well-defined lesion within the left parotid. The tumour is hyperdense compared to the normal parotid tissue, and shows peripheral enhancement. (Reproduced by courtesy of Dr D Montgomery.)

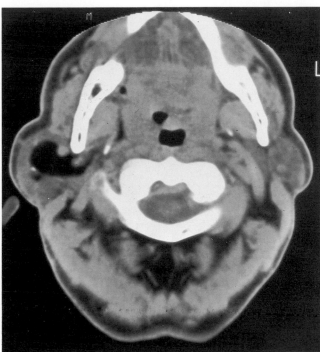

Fig. 6.50 Lipoma of the parotid gland. Transaxial CT scan demonstrating a low-density lipoma within the right parotid. (Reproduced by courtesy of Dr D Montgomery.)

On radionuclide imaging Warthin's tumours appear as areas of increased activity on technetium scans. The activity does not 'wash out' after a sialogogue.

Lipomas

These are uncommon tumours but readily recognized on CT as low-density lesions with well-defined margins. In younger patients they may be admixed with lymphangiomatous elements (Fig. 6.50).

Malignant salivary tumours

These occur most commonly in older female patients and prior radiotherapy is a recognized aetiological factor. Patients who suffer from malignant salivary tumours are at increased risk of second primary tumours elsewhere, for example carcinoma of the breast.

The commonest malignancies are mucoepidermoid carcinomas, adenocystic carcinomas and malignant pleomorphic tumours. Lymphoma may also affect the salivary glands.

Clinical features

Parotid tumours present as palpable masses which may be painful. Facial nerve involvement is a strong indication of malignancy. A submandibular tumour typically presents as a mass which may be associated with pain. Advanced lesions may be fixed to the skin or erode through the skin surface. Sublingual tumours often present as an ulcerated mass in the floor of the mouth and may invade the tongue, mandible or submental soft tissue. Tumours of the minor salivary glands arise as non-ulcerating masses on the hard palate or base of tongue. Lesions of the palate nearly always involve its lateral aspect.

Imaging

As discussed above, the characteristics of malignant lesions include the presence of ill-defined margins, necrosis, ductal obstruction, local invasion and lymphadenopathy (Figs 6.51 and 6.52). The findings are non-specific and can be seen in inflammatory conditions and in some benign lesions complicated by ductal obstruction and inflammation. Conversely, malignant tumours may show characteristics normally associated with benign lesions (Fig. 6.53).

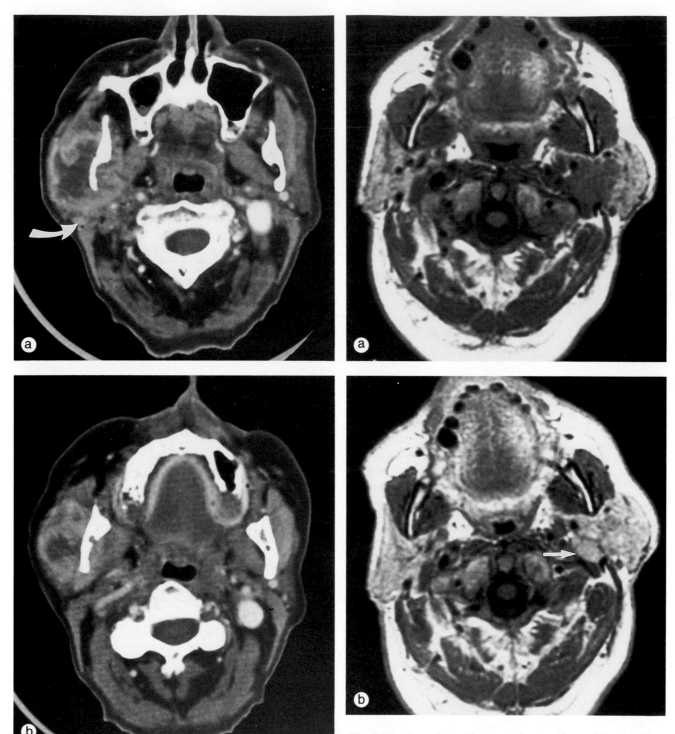

Fig. 6.51 Carcinoma of the parotid. Transaxial postcontrast CT scans (**a**) at the level of the lower maxillary antra; (**b**) at the level of the superior alveolus. A large, poorly defined, necrotic, peripherally enhancing mass involves the superficial portion of the parotid and extends into the deep lobe, displacing the retromandibular vein posterolaterally (arrow in (**a**)) and involving the prestyloid parapharyngeal space with infiltration of the inferolateral portion of the medial pterygoid muscle.

Fig. 6.52 Parotid malignancy in a patient with a facial nerve palsy. (**a**) Precontrast transaxial T1-weighted MR image. (**b**) Postcontrast transaxial T1-weighted MR image. There is a large, medium-signal-intensity mass within the deep lobe of the left parotid gland with extension into the parapharyngeal compartment. The margin between the mass and the normal parotid is ill defined. The tumour shows enhancement after gadolinium DTPA rendering it difficult to separate from the normal parotid, but extension into the posterior belly of the digastric muscle (arrow) is more readily appreciated.

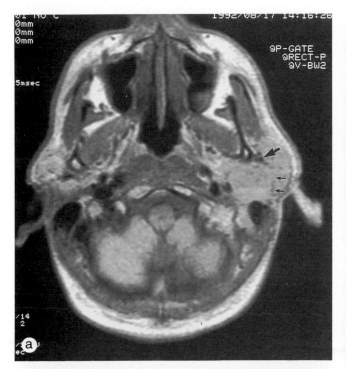

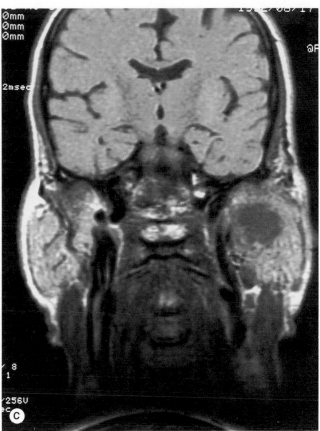

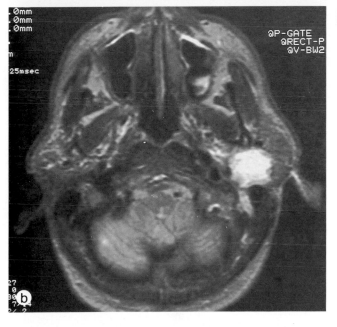

Fig. 6.53 Parotid tumour. Axial proton density (**a**) and T2-weighted (**b**) images; coronal T1-weighted image (**c**). There is a fairly well-defined mass in the deep lobe of the left parotid gland but with some loss of definition around the margins. An initial diagnosis of a locally aggressive but relatively benign lesion was made. Biopsy revealed anaplastic carcinoma, presumably within an intraparotid lymph node. In (**a**) the mass is isointense but displacement of the retromandibular vein (black arrow) and the facial nerve (small black arrows) are well seen. The mass itself is better visualized on the T1- (low signal) and T2-weighted (high signal) images. (Reproduced by courtesy of Dr JPR Jenkins.)

On MRI an attempt has been made to correlate parotid tumour signal intensity with histological grade. It was thought that low-grade malignant tumours behaved in a similar fashion to benign lesions, that is, they were of low signal intensity on T1-weighted images and high signal intensity on T2-weighted images [30]. Conversely, high-grade parotid tumours were supposed to exhibit low signal intensity on both T1- and T2-weighted images [30]. However, experience with larger patient numbers has shown that high-grade tumours may be low or high signal intensity on T2-weighted sequences [31]. This underlies the need for histological confirmation of tumour type in all cases [32].

Mucoepidermoid carcinoma

These are the commonest malignant tumours of the parotid gland, accounting for between 5% and 10% of all tumours. Histologically they are composed of mucous, epidermoid, clear and glandular cells in varying proportions. They are usually classified as high-, intermediate- or low-grade malignancy.

Low-grade lesions have a relatively slow growth rate and may appear benign on imaging. On imaging they appear similar to benign pleomorphic adenomas, but they can show a pattern of diffuse infiltration as seen in high-grade lesions.

The high-grade lesions characteristically infiltrate and destroy the parenchyma, and major ducts can become obstructed. On enhanced scans the appearances are variable.

Adenocystic carcinoma

These [31] are the commonest malignant lesions of the submandibular gland and minor salivary glands, accounting for two-thirds of the tumours occurring in the latter. Histologically the tumours are composed of small uniform cells arranged in cords or nests and characteristically these lesions infiltrate the adjacent tissues and have a propensity for perineural invasion. They are usually slow-growing lesions with metastases occurring up to several years after surgical removal of the primary. Local recurrence occurs in more than 50% of cases.

On imaging they appear as well-defined lesions when small but become less well defined as they enlarge because of invasion of adjacent tissues and an associated fibroblastic reaction.

Lymphomas

Lymphoma may involve the intra- and periglandular lymph nodes and/or the parenchyma. Parenchymal involvement is uncommon, accounting for less than 1% of primary lymphomas and approximately 5% of all extranodal lymphomas. It is commoner in non-Hodgkin's lymphoma than in Hodgkin's disease and the parotid gland is the site of predilection. Overall, the lymphomas account for approximately 3% of parotid tumours, the majority being due to intra-parotid nodal involvement (Fig. 6.54). There is an association with autoimmune diseases, particularly Sjögren's syndrome. On imaging, primary lymphoma (Fig. 6.55) is characterized by a dense infiltrative process that replaces the parenchyma. Lymphadenopathy may or may not be present.

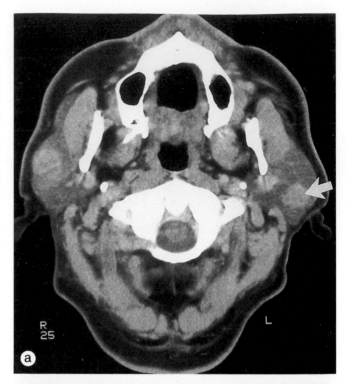

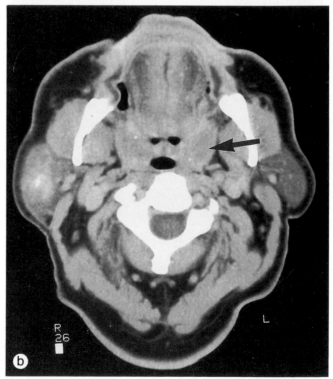

Fig. 6.54 Lymphoma of the intraparotid lymph glands.
Transaxial postcontrast CT scans through the parotid glands demonstrating bilateral intraparotid lymph node involvement (short white arrow in (**a**)) and symmetrical involvement of the tonsils (black arrow in (**b**)).

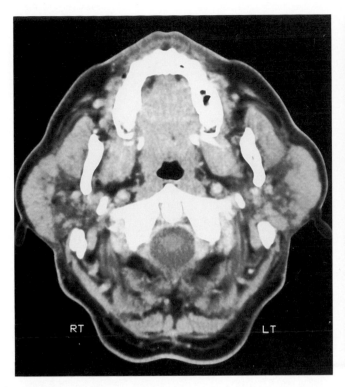

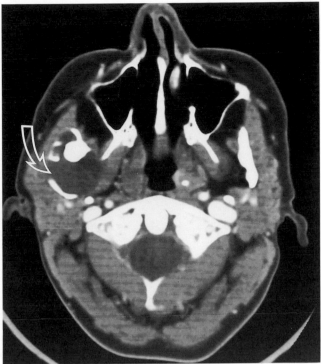

Fig. 6.55 Parotid lymphoma. Transaxial postcontrast CT scan showing parenchymal parotid involvement with only a little normal residual parotid tissue.

Fig. 6.56 Odontogenic keratocyst in a patient with Gorlin's syndrome. Transaxial postcontrast CT image showing an expansile cystic lesion (curved white arrow) in the right mandibular ramus.

THE JAW

Jaw cysts

Jaw cysts usually arise from a proliferation of epithelial nests remaining within the bone after tooth development. They are classified as odontogenic or non-odontogenic depending on whether they are found in tooth-bearing areas of the jaw. They include the following:

1. **Periodontal cysts**, which account for two-thirds of all jaw cysts and develop secondary to infection in a root canal. Seventy-five per cent of periodontal cysts occur in the maxilla and they are found more commonly in adults than in children.
2. **Dentigerous cysts**, which surround the crown of a non-erupted or partially erupted tooth and account for 15% of all jaw cysts.
3. **Odontogenic keratocysts** (primordial cysts), which account for 5% of jaw cysts and are typically seen in young adult men. They also occur in association with Gorlin's basal cell naevus syndrome (Fig. 6.56). Odontogenic keratocysts arise from remnants of the dental lamina and often appear multiloculated.

4. **Nasopalatine cysts**. These are rare and form in the midline anterior maxilla, probably from the epithelium of the nasopalatine ducts in the incisor canal (Fig. 6.57).

The jaw may also be affected by metabolic bone disease for example rickets and Paget's disease, or dysplasias such as fibrous dysplasia, osteogenesis imperfecta, achondroplasia, and cleidocranial dysplasia.

Tumours of the jaw

Benign tumours of the jaw include fibromas, chondromas, osteomas, exostoses and giant cell tumours. Ameloblastoma is a benign but locally invasive tumour which originates in the odontogenic epithelium and presents in middle-aged patients. Eighty per cent occur in the mandibular ramus and give a honeycomb, expansile appearance on plain films.

Malignant tumours of the jaw are rare and are most frequently osteosarcomas which typically occur in older patients suffering from Paget's disease. Other malignant lesions occasionally found in the jaw include non-Hodgkin's lymphoma and metastases.

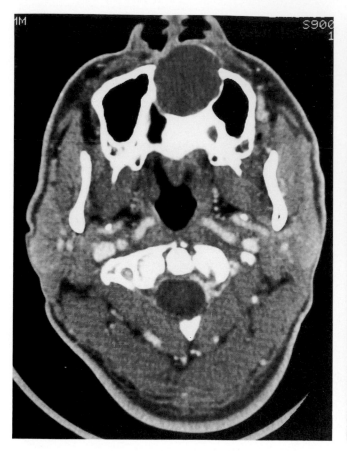

Fig. 6.57 Nasopalatine duct cyst. Transaxial postcontrast CT image. The nasopalatine duct cyst is seen as a well-defined midline lesion.

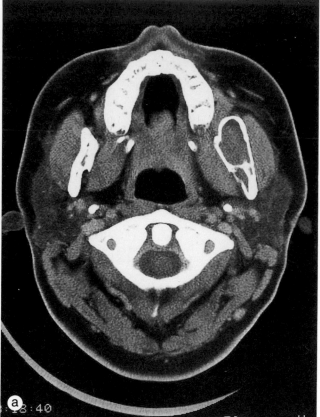

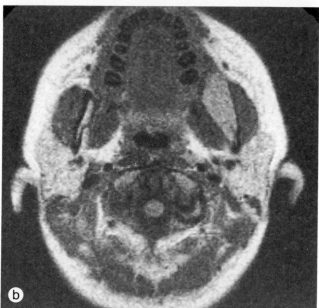

Fig. 6.58 Haemorrhagic mandibular bone cyst. (**a**) Transaxial postcontrast CT image. (**b**) Transaxial T1-weighted MR image. (**c**) Coronal T2-weighted MR image. There is a cystic lesion seen in the left mandibular ramus, expanding but not destroying the bone. High signal intensity on both T1- and T2-weighted images is consistent with haemorrhage. A haemorrhagic simple bone cyst was found at resection.

CT and MRI appearances

Cystic jaw lesions show up as well defined low density masses on CT. On MRI, the cyst fluid content typically causes low signal intensity on T1-weighted images and high signal intensity on T2-weighted images. Haemorrhage occurring within the lesion results in high signal intensity on both T1- and T2-weighted images (Fig. 6.58).

In malignant bone-forming tumours, CT is the preferred investigation for identifying the extent of bone destruction or neoplastic new bone formation, but MRI is superior in assessing the intramedullary extent of disease and infiltration of adjacent organs and tissues.

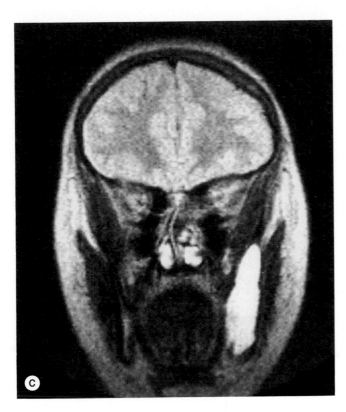

ACKNOWLEDGMENTS

We would like to thank Mr D. Ellard for photographic assistance and Mrs K. Ramnarain and Mrs K. Westwell for typing the manuscript.

REFERENCES

1. Williams PL, Warwick R, Dyson M, Bannister LH, eds. Gray's anatomy. 37th ed. Edinburgh: Churchill Livingstone, 1989.
2. Netter FH. Atlas of Human Anatomy. New Jersey: Ciba-Geigy, 1989.
3. Teresi LM, Lufkin RB et al. MR imaging of the nasopharynx and floor of the middle cranial fossa. Part I. Normal anatomy. Radiology 1987; 164: 811.
4. Cheung YK, Sham JST, Chan FL, Leong LLY, Choy D. Computed tomography of paranasopharyngeal spaces: normal variations and criteria for tumour extension. Clinical Radiology 1992; 45: 109–113.
5. Stutley J, Cooke J, Parsons C. Normal CT anatomy of the tongue, floor of mouth and oropharynx. Clinical Radiology 1989; 40: 248–253.
6. Kassel EE, Keller MA, Kucharczyk W. MRI of the floor of the mouth, tongue and orohypopharynx. Radiol Clin N Amer 1989; 27: 331–351.
7. Cross RR, Shapiro MD, Som PM. MRI of the parapharyngeal space. Radiol Clin N Amer 1989; 27: 353–378.
8. Delbalso AM, Pruet CW, Heffner DK, Carlin MS. The parapharyngeal and retropharyngeal spaces. In: Delbalso AM, ed. Maxillofacial imaging. Philadelphia: WB Saunders, 1990: p 247–261.
9. Delbalso AM, Heffner DK, Pruet CW, Carlin MS. The nasopharynx and oropharynx. In: Delbalso AM, ed. Maxillofacial imaging. Philadelphia: WB Saunders 1990: p 215–219.
10. Mancuso AA, Som PM. The upper aerodigestive tract (nasopharynx, oropharynx and floor of mouth). In: Bergeron RT, Osborn AG, Som PM, eds. Head and neck imaging excluding the brain. St Louis: CV Mosby 1980: p 374–401.
11. Som PM, Cohen BA, Sacher M, et al. The angiomatous polyp and angiofibroma: two different lesions. Radiology 1982: 144: 329–334.
12. Mancuso AA, Hanafee WN. Computed tomography and magnetic resonance imaging of the head and neck. 2nd ed. Baltimore: Williams and Wilkins, 1985: p 440–442.
13. Bohman LG, Mancuso AA, Thomason J, Hanafee WN. CT approach to benign nasopharyngeal masses, American Journal of Roentgenology 1981; 136: 173–180.
14. Fishman EK, Pakter RL, Gayler BW, Wheeler PS, Siegelman SS. Jugular vein thrombosis: diagnosis by computed tomography. J Comput Assist Tomography 1984; 8: 963–968.
15. Million RR, Cassisi NJ, Clark JR. Cancer of the head and neck. In: DeVita VT, Hellman S, Rosenberg SA, eds. Cancer: principles and practice of oncology. Washington, DC: JB Lippincott, 1989: p 488–590.
16. Robinson JD, Crawford SC, Teresi LM et al. Extracranial lesions of the head and neck: preliminary experience with Gd-DTPA-enhanced MR imaging. Radiology 1989; 172: 165–170.
17. Crawford SC, Harnsberger HR, Lufkin RB et al. The role of gadolinium DTPA in the evaluation of extracranial head and neck mass lesions. Radiol Clin N Amer 1989; 27: 219–242.
18. Laine FJ, Braun IF, Jensen ME, Nadel L, Som PM. Perineural tumour extension through the foramen ovale: evaluation with MR imaging. Radiology 1990; 174: 65–71.
19. van den Brekel MWM, Stel HV, Castelijns JA, Nauta JJP, van der Waal I, Valk J, Meyer CJLM, Snow GB. Cervical lymph node metastasis: Assessment of radiologic criteria. Radiology 1990; 177: 379–384.
20. Yousem DM, Som PM, Hackney DB, Schwaibold F, Hendrix RA. Central nodal necrosis and extracapsular neoplastic spread in cervical lymph nodes: MR imaging versus CT. Radiology 1992; 182: 753–759.
21. Teresi LM, Lufkin RB et al. MR imaging of the nasopharynx and floor of the middle cranial fossa.

Part II. Malignant tumours, Radiology 1987; 164: 817–821.

22. Willinsky R, Kassel EE, Cooper PW, Chin Sang HB, Haight J. Computed tomography of the lingual thyroid. J Comput Assist Tomography 1987; 11: 182–183.

23. Maran AGD. Benign disease of the neck. In: Brown W, Scott G, eds. Otolaryngology, vol V: 5th ed. Stell PM (ed) Laryngology. Oxford: Butterworths, 1987, p 283–300

24. De Schepper AM, Monheim P, Degryse HR, Van de Heyning P. CT of second branchial cleft cysts and fistula: comparison with MRI in three cases. Ann Radiol 1988; 31: 141–148.

25. Cooke J, Parsons C. Computed tomographic scanning in patients with carcinoma of the tongue. Clin Radiol 1989; 40: 254–256.

26. Cooke J, Morrison G, Keyserlingk J, Parsons C. CT scanning in patients following surgery on the tongue and floor of mouth. Clin Radiol 1990; 41: 306–311.

27. Sugimura K, Kuroda S, Furukawa T, Matsuda S, Yoshimura Y, Ishida T. Tongue cancer treated with irradiation: assessment with MR imaging. Clin Radiol 1992; 46: 243–247.

28. Tabor EK, Curtin HD. MR of the salivary glands. Radiol Clin N Amer 1989; 27: 379–392.

29. Swartz JD, Rothman MI, Marlowe FI et al. MR imaging for parotid mass lesions: attempts at histopathologic differentiation. J Comput Assist Tomography 1989; 13: 789–796.

30. Som PM, Biller HF. High-grade malignancies of the parotid gland; identification with MR imaging. Radiology 1989; 173: 823–826.

31. Sigal R, Monnet O, de Baere T, Michaeu C, Shapeero LG, Julieron M et al. Adenoid cystic carcinoma of the head and neck: evaluation with MR imaging and clinical-pathologic correlation in 27 patients. Radiology 1992; 184: 95–101.

32. Yousem DM. Dashed hopes for MR imaging of the head and neck: the power of the needle. Radiology 1992; 184: 25–26.

7 The larynx and hypopharynx

Stephen J. Golding

INTRODUCTION

This chapter considers the modern imaging of the larynx and hypopharynx, i.e. the airway and gastrointestinal tract between the hyoid bone superiorly and the first tracheal ring and the oesophagus inferiorly. This is an area of anatomical complexity and therefore well suited to modern planar imaging techniques, with their ability to display anatomy in sections and to discriminate between tissues. Indeed, computed tomography (CT) and magnetic resonance imaging (MRI) have opened up for the general radiologist this area, which was formerly the province of the specialist. For these reasons CT and MRI have to a large extent replaced conventional imaging techniques in clinical practice [1].

THE CLINICAL ROLE OF IMAGING

Radiological investigation is required in only a minority of patients with pathology in this area. Many of the most common problems, for example laryngitis or pharyngitis, are treated on the basis of clinical diagnosis alone. The throat is also an area which is amenable to direct inspection or to indirect pharyngoscopy/laryngoscopy or endoscopy, which allow clinical inspection of the lining mucosa. Where doubt exists, biopsy can be carried out under direct vision. Imaging is required only when these simpler techniques are an insufficient basis for clinical management.

Imaging investigations are of two types, dynamic and static. Dynamic imaging is usually performed by barium swallow examination but may also be carried out by scintigraphy or ultrasound. These techniques are used to detect defects in function of the pharynx, or to reveal lesions which may only be detectable at a particular phase of the swallowing cycle, or to guide treatment in patients with swallowing problems. This technique is beyond the scope of the present chapter and the reader is referred to the comprehensive account by Jones and Donner [2].

Static imaging of this area is now heavily dependent on CT and MRI, and to a lesser extent ultrasound. Although the range of pathology which arises from this area is wide, symptomatically it is convenient to consider the larynx and hypopharynx together.

The main diagnostic role of imaging is the investigation of a suspected mass. Non-specific symptoms may be present, such as dysphagia, stridor or other respiratory problems, or enlargement of the neck. More specific symptoms may point to a precise anatomical site; for example change of voice in tumours of the vocal fold, or pain at a specific point. Examination of this area is also carried out in patients presenting with cervical lymphadenopathy of unapparent origin. Recurrent infection may be a pointer to a congenital lesion such as a laryngocele and occasionally pharyngeal disease may present with pain in the ear, referred through the vagus nerve from involvement of the superior laryngeal nerve.

Imaging has an important role in the investigation of the extent of known masses, i.e. disease staging. This applies not only to malignant tumours prior to radical treatment, but also to the resection of benign lesions. Evaluation of extension of disease outside the clinically amenable areas is the principal value of investigation. Imaging is also required to monitor treatment of unresected lesions and to detect recurrent disease.

RELATIONSHIP OF CT AND MRI TO OTHER IMAGING TECHNIQUES

Conventional radiographs – usually frontal and lateral views – outline the structures of this area by contrast between the air column and surrounding soft tissues. They also demonstrate the laryngeal cartilages when these are calcified or ossified. Conventional radiographs are indispensable for the management of acute trauma, including the demonstration of opaque foreign bodies. They are also used to assess airway compression, or to demonstrate abnormal air patterns such as occur in laryngoceles. Disease which erodes cartilage can sometimes be inferred if the cartilages are heavily calcified but otherwise radiography, including conventional tomography, has limited soft tissue resolution and for other applications has been replaced by CT and MRI [3].

In the past conventional tomography was widely used to assess the anatomy and function of the vocal folds; this has largely been replaced by clinical evaluation under indirect laryngoscopy and investigation by CT or MRI.

Procedures using contrast media are also employed in this area. Conventional barium swallow is valuable in dysphagic patients, when it may show pharyngeal compression. Barium-coated mucosal views may also be used to assess smaller lesions involving the mucosa, but indirect pharyngoscopy is more reliable and more likely to be used [4]. As mentioned previously, barium studies have an important role in the dynamic assessment of the pharynx.

Mucosal coating of the lining of the larynx by

contrast medium – laryngography [3] – is also capable of showing small mucosal lesions but shares with barium swallow an inability to demonstrate the deep extent of lesions, or lesions which do not involve the mucosa. It is now practised infrequently.

Ultrasound has been recommended for the diagnosis of masses in the neck, having the advantages that it does not use ionizing radiation and that some tissue characterization is possible, particularly between cysts and soft tissue masses. However in this area this distinction may not be as reliable as elsewhere in the body, as cysts frequently have contents which may produce an echo pattern similar to that of soft tissues. Ultrasound is also limited by reflection of sound at air/tissue interfaces and is therefore of limited value in evaluating lesions near the airway but nevertheless has been recommended for some laryngeal lesions [3,5]. Ultrasound is a valuable method of diagnosis where expertise in cervicofacial examination is available; in general it has not displaced CT or MRI but it is likely that its applications will grow with increasing experience.

EXAMINATION TECHNIQUE

The main principles have been covered in Chapter 1. However in both CT and MRI there are specific technical requirements for the examination of the larynx and hypopharynx [6].

CT

CT examination is limited to the axial plane. In general sections should be planned from the lateral localizer view as this shows the important landmarks – hyoid, epiglottis, laryngeal ventricle (Fig. 7.1). The structures of the larynx are small and relatively thin sections are required. Two techniques exist:

1. contiguous sections 4–5 mm thickness, which provide reasonably good signal-to-noise ratio and soft tissue contrast. The spacing depends on the size of pathology; larger lesions can be examined with sections at 10 mm intervals, others require contiguous sections [7];
2. fine [1–2 mm] contiguous sections taken through the area. Thinner sections have a poorer signal-to-noise ratio but multiplanar reconstruction is possible from this technique [8].

In either case, other sections may be required outside the primary field of interest, for example for

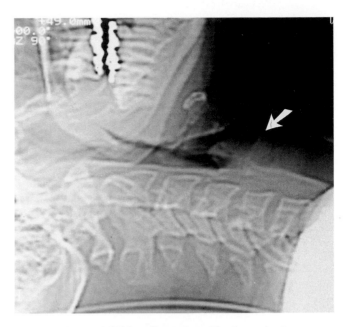

Fig. 7.1 Lateral CT localizer view. The important landmarks for planning the examination are the epiglottis, hyoid bone and vocal fold (arrow), which can be identified by contrast with the laryngeal ventricle above it. The gantry may be angled so that sections are parallel to the line of the vocal fold.

the evaluation of lymphatic drainage areas in the case of staging primary tumours. In these circumstances sections should be obtained from the floor of the mouth to the root of the neck. When tumours are being staged it is important that the referral details state the precise site of the lesion, as contiguous sections will be needed through this area, even if non-contiguous sections are being used to detect lymphadenopathy.

The field of view should be small – 20 cm is appropriate – in order to provide good spatial resolution. A soft tissue algorithm is used for image construction. When CT is carried out for diagnostic purposes it is preferable to reduce the absorbed dose as far as possible by using a low mAs, provided this does not adversely affect image quality. Avoiding the use of contiguous sections, where possible, also reduces the absorbed dose to sensitive structures such as the thyroid gland.

Contrast medium enhancement is frequently used in diagnostic examinations and is essential for those carried out for disease staging. Enhancement distinguishes perfused lesions (i.e. those with a circulation and therefore consisting of tissue) and unperfused lesions such as cysts, abscesses and necrotic tissue. Enhancement also improves soft tissue definition in this area and allows enlarged lymph nodes to be readily distinguished from surrounding blood vessels (Fig. 7.2).

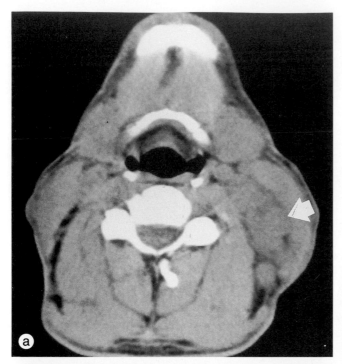

Problems in CT include movement artefact due to swallowing. It is usual to ask the patient to breathe quietly during the exposure. Pooling of saliva, which may produce diagnostic pitfalls in the valleculae and pyriform fossae may be avoided by asking the patient to swallow between exposures. This problem may also be overcome by asking the patient to open his/her mouth or to phonate during the exposure. This opens the valleculae and pyriform fossae. Valsava manoeuvre of the glottis should be avoided during breath-holding because glottal closure distorts the appearances. However an oral Valsava manoeuvre is sometimes valuable in opening the pyriform fossa if the measures described above fail (Fig. 7.3).

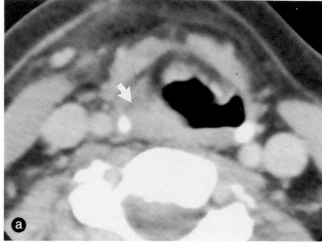

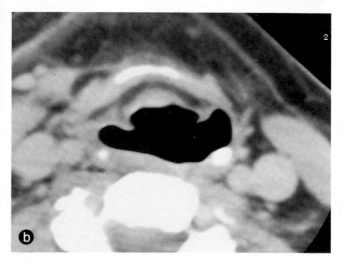

Fig. 7.2 The value of contrast enhancement in CT. In unenhanced section (**a**) there is asymmetry of the structures lying beneath the left sternomastoid muscle (arrow); this suggests lymphadenopathy but the vessels at this level can be very variable. The image after enhancement (**b**) shows intense contrast medium uptake in blood vessels and the submandibular glands (arrows) but in addition there is poorly enhancing lymphadenopathy (L) on the left side.

Fig. 7.3 Difficulty in interpretation in the pyriform fossae on CT. An initial section (**a**) appears to show a mass (arrow) occupying the right pyriform fossa. Similar appearances were obtained with the mouth open. However when the patient was asked to perform an oral Valsalva manoeuvre the pyriform fossa opened (**b**). Apparent thickening of the wall of the fossa is due to it being only partly distended; endoscopic findings were normal.

MRI

This has the advantage over CT of superior soft tissue contrast and the ability to image in multiple planes.

High resolution examination requires the use of close-coupled coils. Posterior coils for the examination of the cervical spine are generally available and may yield adequate signal intensity from structures at the front of the neck. Proprietary anterior neck coils are available although these are tolerated poorly by some patients and are susceptible to movement artefact. Harnessed coils which produce a uniform imaging volume across the neck usually provide the best image quality [9].

Most standard examinations employ the spin echo sequence and include a coronal or sagittal T1-weighted sequence, supplemented by axial T1-weighted images and sometimes also axial T2-weighted images [6]. This choice depends on the clinical indication and is covered in detail below.

As in CT, fine sections are advisable; the author's own technique is to use 4 mm sections with a 1 mm interslice gap. A small field of view should be used, providing this does not cause aliasing artefact. The display matrix should be fine if possible, especially when examining the vocal folds, although this decreases the signal-to-noise ratio.

Problems in MRI usually stem from movement artefact due to swallowing during the acquisition sequence. This is a particular problem in patients with pharyngolaryngeal pathology, as salivary secretion or discharge may be copious. In this situation a faster examination, including individual sections during suspended respiration if the technology is available, may be obtained using a gradient echo technique. Although this has less appropriate contrast resolution than the spin-echo technique, diagnostic information may be obtained without artefact (Fig. 7.4).

In many instances CT and MRI have a complementary role in the investigation of the larynx and hypopharynx, CT currently having superior spatial resolution, whereas MRI has the better contrast resolution of different tissues.

ANATOMY

The anatomy of this area is complex but in outline is as follows. The skeletal elements, from superior to inferior, consist of the arch of the hyoid bone, the paired wings of the thyroid cartilage and the cricoid cartilage. These are linked by fascial membranes anteriorly and laterally but are open posteriorly,

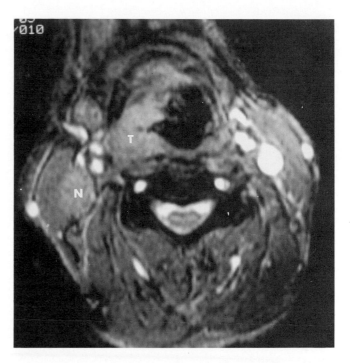

Fig. 7.4 Use of gradient echo technique in MRI. In this patient with an extensive tumour (T) of the right pyriform fossa, acceptable images could not be obtained with the spin echo technique because of pharyngeal discharge and movement artefact due to swallowing. Using a T2-weighted gradient echo sequence, 10 images were obtained in 1 minute 20 seconds, without undue artefact. Note that this technique produces high signal from flowing blood and identifies a pathologically enlarged lymph node (N), beneath the right sternomatoid muscle.

except the cricoid cartilage, which is a complete ring. The paired arytenoid cartilages sit on the superior aspect of the cricoid cartilage posteriorly. The flap-like structure of the epiglottis rises from inside the junction of the thyroid cartilages anteriorly, projecting obliquely in a supero-posterior direction.

Anteriorly and laterally the laryngeal skeleton is also connected by strap muscles which suspend the larynx from the floor of the mouth and elevate the larynx during swallowing. The inferior pharyngeal constrictor muscle arises from both the thyroid and cricoid cartilages and provides the posterior wall, enclosing the hypopharynx. Its inferior fibres blend with those of the upper oesophagus, creating the cricopharyngeal sphincter posterior to the cricoid ring.

The components of the laryngeal skeleton are also linked by small intrinsic muscles which are responsible for the internal movements of the larynx. They are not well seen at imaging, except for the thyro-arytenoid muscle and the vocalis muscle which runs from the arytenoid cartilage on each side to the thyroid junction, forming the glottis.

The area is lined with squamous mucosa which is drawn up into several important folds. From the posterior aspect of the tongue to the anterior surface of the epiglottis a midline frenulum divides the pre-epiglottic space into the two valleculae. The aryepiglottic folds run obliquely downwards and backwards from the lateral surface of the epiglottis, gradually separating the hypopharyngeal airway into larynx and paired lateral spaces – the pyriform fossae – which lead from the valleculae to the cricopharyngeal sphincter around the laryngeal vestibule.

Within the larnxy the mucosa extends over the vocalis muscle, creating the true focal fold. A short distance above this the mucosa forms the less prominent false vocal fold, separated from the true fold by a small air-containing space – the laryngeal ventricle – which reaches the outer wall in between the two folds. Below the true vocal fold the mucosa runs inferiorly over the conus elasticus, which slopes downwards and outwards to the lip of the cricoid cartilage.

The laryngeal mucosa is also separated from its muscles and skeleton by fat and loose areolar tissue, with two important concentrations. Firstly, immediately inferior to the valleculae fat occupies the space between the epiglottis and the hyoid cartilage anteriorly. Secondly, the false vocal fold largely contains fat. These are important imaging landmarks in view of the ease of characterizing fat on both CT and MRI.

The normal appearances on both CT and MRI are shown in Figs 7.5–7.7. Tissue characterization naturally differs between the techniques but elements are well characterized by both [1]. Definition of cartilage is usually best with CT, as in the adult it usually contains sufficient calcium to be of high attenuation. Owing to the greater sensitivity of CT compared to conventional radiography, variations in the calcification pattern are not so much a diagnostic problem in CT as in radiography. A variable amount of fatty marrow is present in the cartilages, especially the thyroid and cricoid, and this appears as a low-attenuation medullary zone on CT.

Fig. 7.5 Normal appearances of axial CT and T1-weighted spin echo MRI. The series consists of axial sections from just below the free portion of the epiglottis down to the trachea. Sections were obtained parallel to the vocal fold. The CT and MRI studies were obtained from different subjects and some details of the anatomy do not correspond entirely. Note that the cervical blood vessels are not clearly identified on CT, except in figures (i) and (j), where enhancement was used. On MRI the vessels are easily identified because on the spin echo sequence flowing blood produces a signal void.

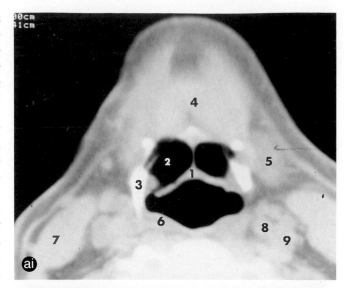

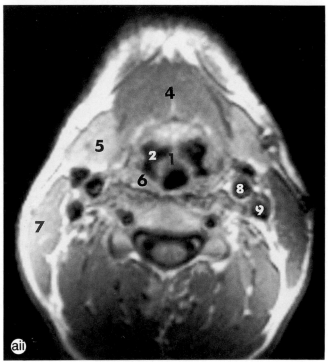

(a) Section through the valleculae. A midline frenulum from the epiglottis to the anterior wall divides the paired valleculae. This section also shows the cornua of the hyoid bone. Note that the epiglottis is in contact with the left lateral wall on CT and the posterior wall on MRI; both these appearances are normal.

1 = epiglottis; 2 = vallecula; 3 = hyoid bone; 4 = myelohyoid muscle; 5 = submandibular gland; 6 = pharyngeal constrictor muscle; 7 = sternomastoid muscle; 8 = carotid artery; 9 = jugular vein; 10 = strap muscles; 11 = aryepiglottic folds; 12 = thyroid cartilage; 13 = pyriform fossa; 14 = false vocal fold; 15 = cricoid cartilage; 16 = true vocal fold; 17 = conus elasticus; 18 = thyroid gland; 19 = cricopharyngeal sphincter.

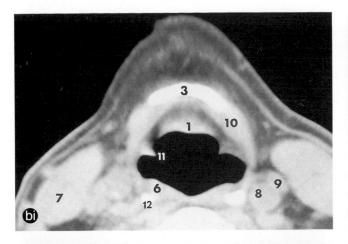

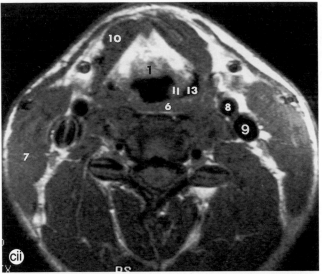

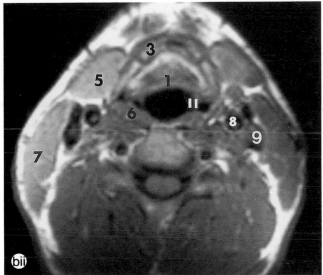

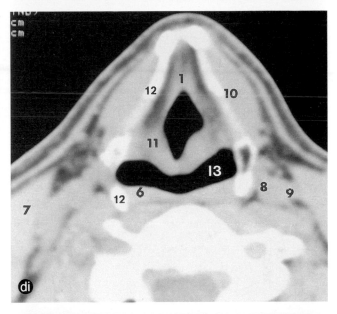

(b) Section through the body of the hyoid bone. The hyoid bone contains little fatty marrow but is identified on MRI by contrast with the surrounding strap muscles. At this level below the inferior limit of the valleculae the space between the hyoid and the epiglottis is occupied by fat. This is the superior limit of the aryepiglottic folds, which arise from the posterior aspect of the epiglottis.

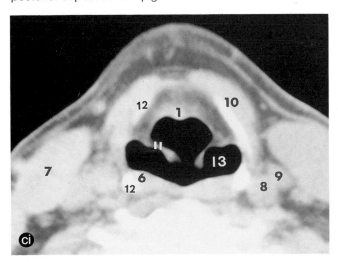

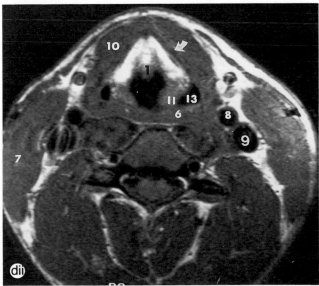

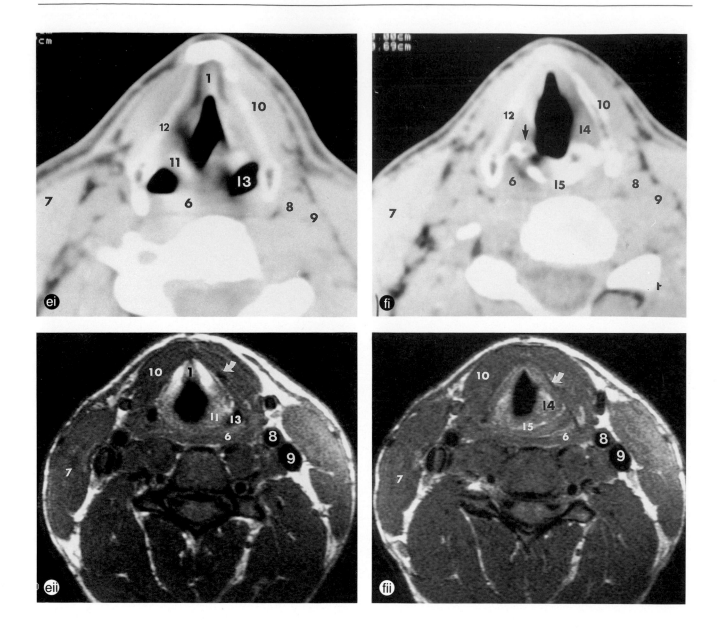

(c)–(e) As the laryngeal vestibule is descended the aryepiglottic folds slope downwards and backwards, coming to meet in the midline posteriorly and divide the pharynx into paired pyriform fossae. The folds thicken as they descend and show increased attenuation on CT, decreased signal intensity on MRI, owing to increasing content of muscle fibres in the inferior part of the laryngeal additus. At this level the lamina of the thyroid cartilage contains little fatty marrow and is difficult to identify on MRI, except by areas of signal void due to calcification (white arrow).

(f) Section through the false vocal folds. This level often includes the superior process of the arytenoid cartilage (black arrow). The false vocal cords contain more fat than the true folds and therefore have a higher signal intensity on MRI and lower attenuation on CT.

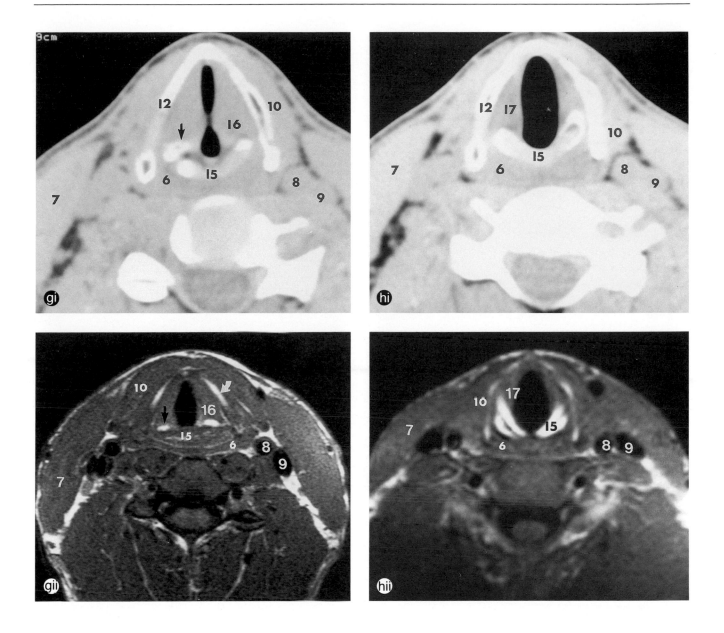

(g) Section through the glottis. The transverse process of the arytenoid cartilage (black arrow) is shown at this level. On CT the subject suspended respiration with partial closure of the glottis and the cords are partially adducted. The MRI study shows the normal resting position of abduction. At this level the thyroid cartilage contains more fatty marrow and is clearly seen on MRI (white arrow).

(h) Section through the conus elasticus. This section shows the inferior limit of the thyroid lamina, with the glottis widening as the walls of the conus slope downwards and outwards to the cricoid ring. The cricoid is tallest posteriorly, so that only the posterior limit is seen at this level.

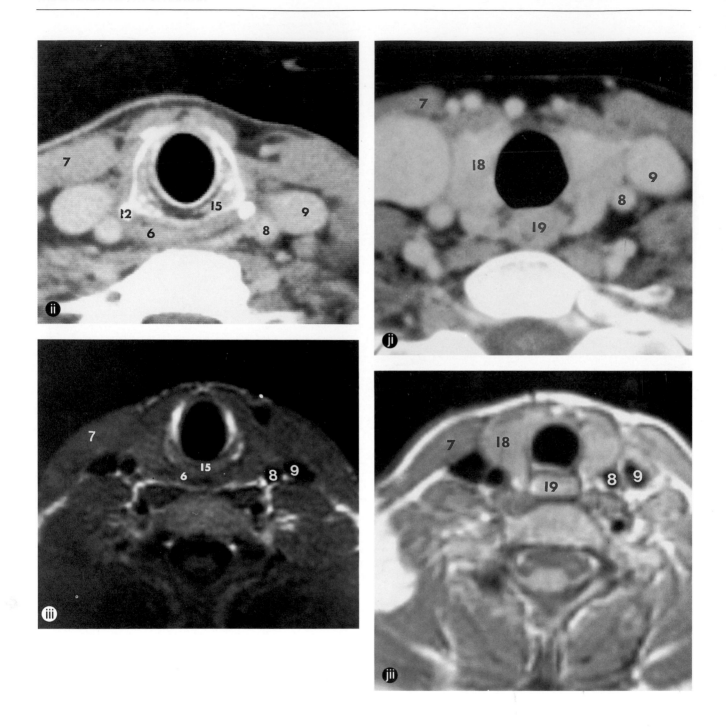

(i)   Section through the cricoid cartilage. The cartilage contains a well-defined marrow cavity on CT, which has a variable fatty component on MRI. Note that the cricoid cortex can be seen as a thin signal void on MRI. At this level below the conus elasticus the mucosa is virtually in contact with the cartilage.

(j)   Section through the thyroid gland. On enhanced CT the gland is of high attenuation. It is also well seen on MRI in this subject but note that in the preceding MRI images, obtained in a different subject, the gland is isointense with cervical muscles. At this level the cricopharyngeal sphincter is well seen.

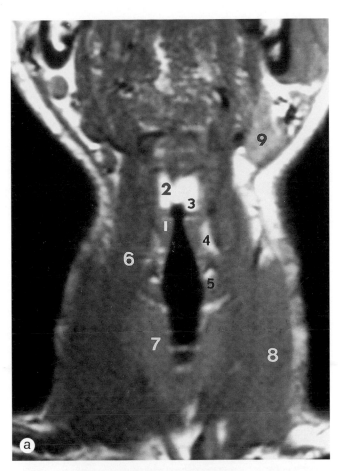

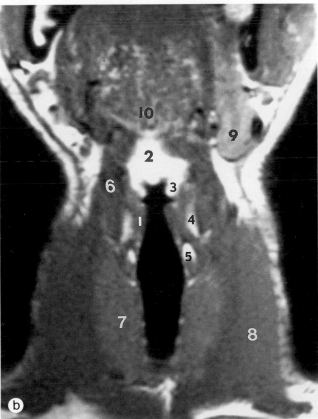

Fig. 7.6 Coronal T1-weighted MRI images.

1 = true vocal fold; 2 = pre-epiglottic fat; 3 = false vocal fold; 4 = thyroid cartilage; 5 = cricoid cartilage; 6 = strap muscles; 7 = thyroid gland; 8 = sternomastoid muscle; 9 = submandibular gland; 10 = myelohyoid muscle, with myoglossus above; 11 = aryepiglottic fold; 12 = pyriform fossa; 13 = cricopharyngeal sphincter.

(a) Through the anterior vocal fold. In this section the intermediate intensity of the thyroarytenoid muscle constitutes most of the vocal fold. Above this the false vocal fold is poorly developed and the fat of the pre-epiglottic space is seen. The cartilages are poorly distinguished from the strap muscles because of low content of fatty marrow and signal void from their cortex.

(b) Through the mid part of the vocal fold. In this image the vocal fold consists of vocalis and thyoarytenoid muscles, with the latter running superiorly towards the thyroid cartilage. The laryngeal ventricle and false vocal cord are well developed at this level.

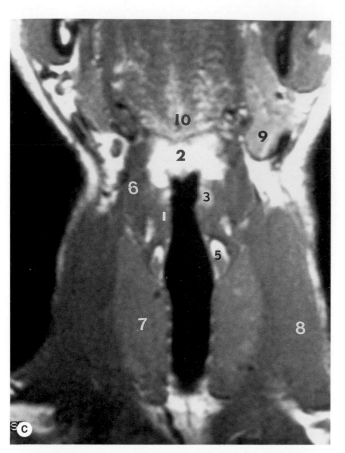

(c) At the posterior limit of the vocal folds. At this point the glottis is at its widest and no ventricle is present. The cricoid cartilage contains more fat and areas of fatty marrow are seen in the thyroid cartilage.

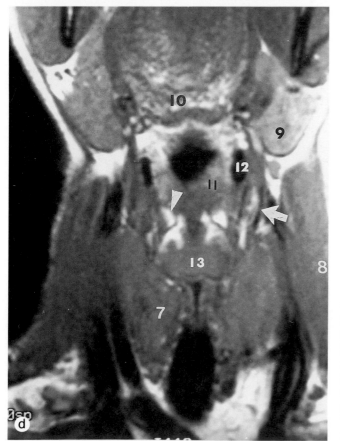

(d) Section through the arytenoid cartilages. The fat of the arytenoid cartilages (arrowhead) can be identified on the superior aspect of the cricoid ring. Above this the inferior, partly muscular, part of the aryepiglottic folds is seen separating the laryngeal vestibule from the pyriform fossae. The normal appearance of thyroid cartilage, combining cortical signal void and mixed attenuation of marrow cavity, is particularly well seen on the left (arrow). This section also includes part of the cricopharyngeal sphincter.

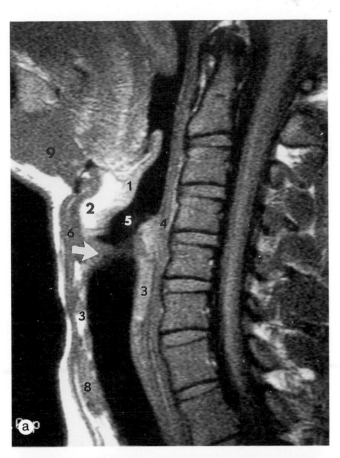

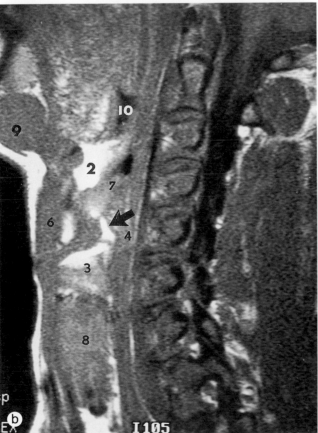

Fig. 7.7 Sagittal T1-weighted MRI images.

1 = epiglottis; 2 = pre-epiglottic fat; 3 = cricoid cartilage; 4 = pharyngeal constrictor muscle; 5 = laryngeal vestibule; 6 = strap muscles; 7 = aryepiglottic fold; 8 = thyroid gland; 9 = myelohyoid muscle; 10 = vallecula.

(a) Midline section. This section has included some of the vocal fold (arrow), particularly anteriorly where the glottis is narrow. The relationship of the epiglottis to the anterior attachment of the false vocal fold is well shown, as is the pre-epiglottic fat space.

(b) Section through the side wall of the larynx shows high signal from fat in the superior limit of the cricoid cartilage and in the vertical process of the arytenoid cartilage (arrow).

On MRI cartilage has a low signal intensity and calcification creates a signal void, whereas fat and marrow have a high signal intensity on T1-weighted spin-echo sequences. Cartilage therefore has a variable appearance on MRI, depending on the relative proportions of calcification and marrow [9]; this particularly affects the hyoid bone, which has little marrow.

The surrounding muscles have intermediate attenuation/signal characteristics on both CT and MRI. As the vocal fold is largely composed of muscle, this has similar characteristics, allowing it to be distinguished from the false vocal fold, which largely contains fat. Fat can also be detected in the aryepiglottic folds on both techniques.

The carotid and jugular vessels are readily distinguished from the strap muscles on spin echo MRI sequences, owing to flow-related signal void (Fig. 7.5) but on CT the blood vessels have a similar attenuation to surrounding soft tissues and are therefore only well seen following enhancement (Fig. 7.2). The thyroid gland is usually of intermediate to high attenuation on CT depending on its iodine content but is of intermediate signal intensity on MRI and may not be distinguishable from the adjacent strap muscles (Fig. 7.5 (i) and (j)).

PATHOLOGY

Although the range of pathology is particularly wide in the head and neck, in clinical practice imaging the larynx and hypopharynx largely concerns malignant disease, congenital lesions such as laryngoceles and branchial cysts, and evaluating the effects of trauma, with the investigation of tumours being the most important and most common application. Investigation is also required for diagnosis of masses arising from surrounding structures.

Many of the principles of investigation of this region are common to all types of neoplasm. In the following account the general approach is described in relation to carcinoma of the larynx.

Squamous carcinoma of the larynx

Squamous carcinoma accounts for the vast majority of laryngeal tumours, is seen in the late middle-aged and elderly, and can be related to cigarette smoking [10]. These patients present commonly with voice change, persistent sore throat or haemoptysis.

Carcinoma of the larynx is classified according to its site of origin, which has prognostic implications [1].

1. Glottic tumours (Figs 7.8–7.13) arise from the vocal folds, including the anterior or posterior commissure. They are usually well differentiated, well localized and metastasize late. This site accounts for approximately two-thirds of all laryngeal carcinomas.

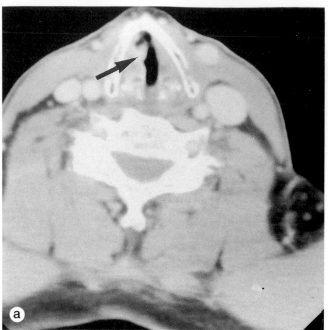

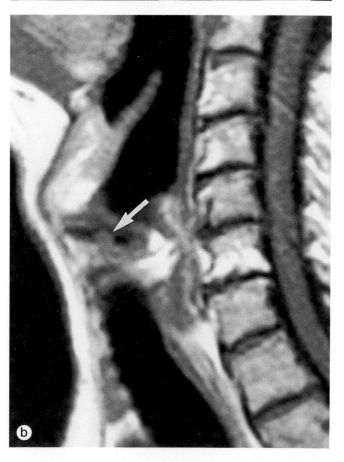

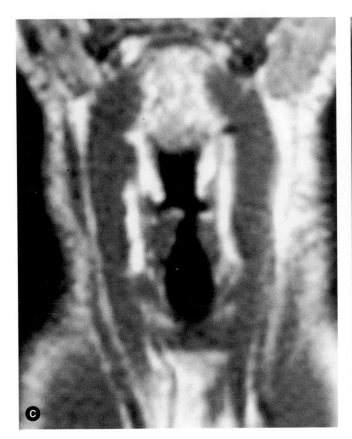

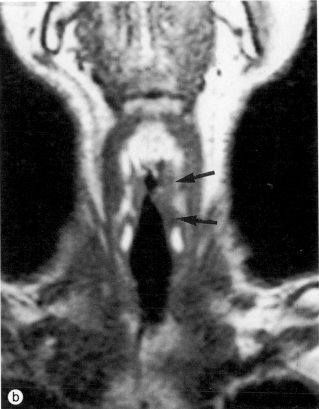

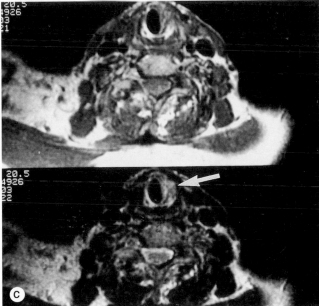

Fig. 7.8 *T*1 **carcinoma of the glottis. (a)** CT showing a nodule arising from the right vocal fold (arrow). **(b)** Sagittal MRI shows the nodule arises from the upper aspect of the vocal fold (arrow). **(c)** MRI in the coronal plane shows expansion of the right vocal fold but no deep infiltration. (MRI images reproduced by kind permission of Dr NR Moore.)

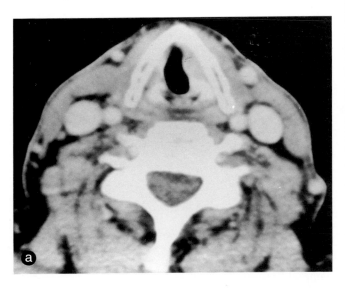

Fig. 7.9 *T*3 **carcinoma of glottis. (a)** CT section showing tumour apparently localized to the left vocal fold. **(b)** Coronal MRI showing expansion of the left vocal fold and also of the thyroarytenoid muscle, both superiorly and inferiorly where the process extends to the cricoid cartilage (arrows). **(c)** Proton density (upper) and T2-weighted (lower) MRI images showing high signal from the tumour, with lateral extension (arrow) into the paralaryngeal fat, although not beyond the larynx. Final radiological staging was *T*3.

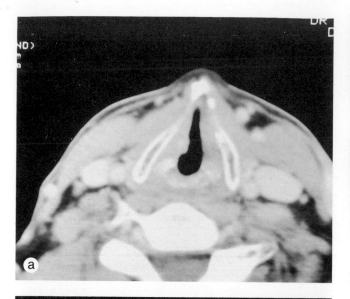

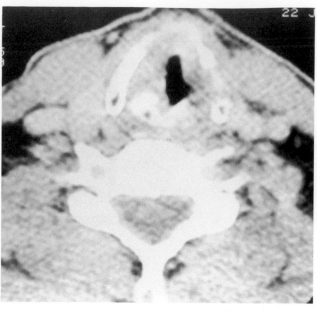

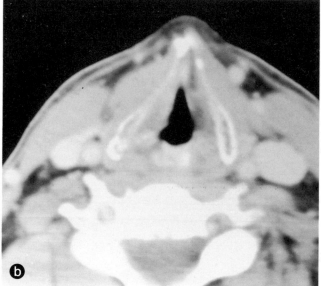

Fig. 7.11   *T*3 **carcinoma of the glottis.** CT sections showing tumour arising in the right vocal fold and crossing the anterior commissure to infiltrate the anterior third of the fold on the left. This precludes vertical hemilaryngectomy.

Fig. 7.10   *T*3 carcinoma of the glottis, supraglottic extension. (**a**) CT section through the vocal folds showing expansion of the right vocal fold. (**b**) CT section through the false vocal folds shows infiltration of the fold on the right, with increased attenuation compared to the normal fat of the left vocal fold.

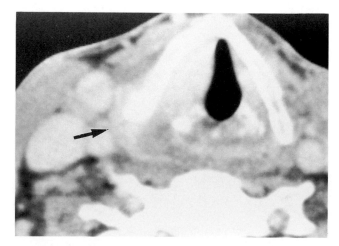

Fig. 7.12   *T*4 **carcinoma of the glottis.** The right vocal fold is expanded by a relatively localized tumour but this enhanced section shows abnormal enhancing tissue extending laterally around the posterior margin of the thyroid cartilage (arrow).

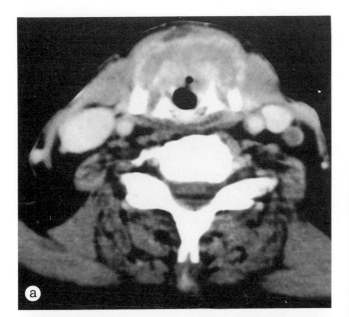

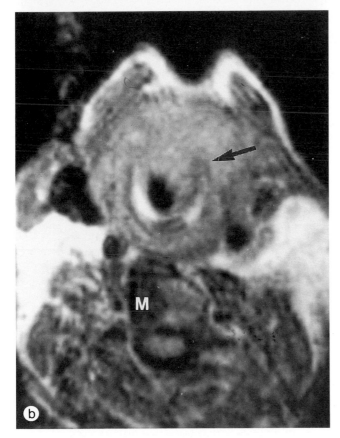

2. Supraglottic tumours (Figs 7.14–7.16) arise from the epiglottis, aryepiglottic folds, false cords or laryngeal ventricles. They tend to be less well differentiated, rapidly growing and poorly defined, with a propensity to metastasize. Because of this they can be difficult to distinguish clinically and radiologically from tumours arising in the hypopharynx and spreading into the larynx (see below).

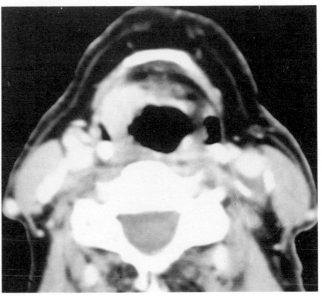

Fig. 7.14   *T*3 supraglottic tumour. CT shows that a carcinoma of the right side of the epiglottis has infiltrated the pre-epiglottic fat.

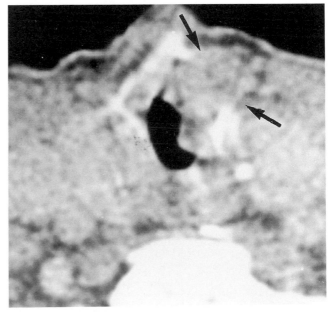

Fig. 7.13   *T*4 carcinoma of the glottis, on CT (**a**) and MRI (**b**). An anterior commissure mass has eroded the thyroid cartilage anteriorly. Note that the margins of tumour are shown by enhancement on CT. On MRI residual fatty marrow in the intact cricoid cartilage facilitates identification of this (arrow). There is also a low signal intensity focus in the right side of the vertebral body on MRI, suggesting metastasis (M).

Fig. 7.15   *T*4 supraglottic tumour. CT shows that this carcinoma of the left false vocal fold has eroded through the thyroid lamina and extends into the overlying tissues (arrows).

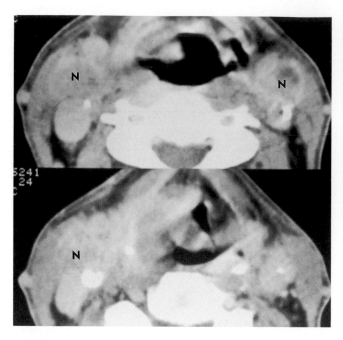

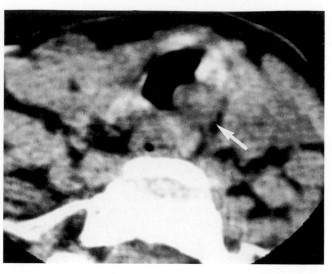

Fig. 7.17 *T*4 subglottic tumour. CT shows that this relatively small subglottic tumour has extended through the laryngeal wall immediately inferior to the cricoid ring and anterior to the cricopharyngeal sphincter (arrow).

Fig. 7.16 *T*4 supraglottic tumour. This CT section shows an extensive tumour involving the epiglottis and aryepiglottic fold on the right side, with contiguous extension into the soft tissues of the neck on the right. There are abnormally enlarged lymph nodes (N) bilaterally.

3. Subglottic tumours (Fig. 7.17) are the least common, and usually least well differentiated. They grow rapidly and present late, arising in a relatively asymptomatic area. They are limited by the conus elasticus and deep invasion is common, with a tendency to early metastasis.

These tumours, with the exception of small sub- glottic tumours immediately beneath the vocal fold, can be diagnosed by indirect laryngoscopy with biopsy if necessary. The main role of imaging is disease staging. *T* staging criteria differ according to the site and are summarized in Table 7.1.

Neither CT nor MRI is reliable in detecting disease localized to the mucosa; a normal examination is therefore consistent with *T*1 disease [11]. Similarly,

---

Table 7.1 *T* staging scheme for carcinoma of the larynx (modified from Reference 21)

Glottic tumours

*T*1   Tumour confined to vocal fold, which is mobile
*T*2   Supraglottic or subglottic extension, and/or impaired vocal fold mobility
*T*3   Tumour limited to the larynx, but with vocal fold fixation
*T*4   Tumour erosion through laryngeal cartilage, or into other surrounding structures

Supraglottic tumours

*T*1   Tumour limited to one subsite, with normal vocal fold mobility
*T*2   More than one subsite involved, or extension to the glottis, with mobile vocal folds
*T*3   Tumour limited to the larynx, with vocal fold fixation, and/or invasion posterior to the larynx, to the pyriform fossa, or to the pre-epiglottic space
*T*4   Tumour extension through laryngeal cartilage, or into other surrounding structures

Subglottic tumours

*T*1   Tumour limited to the subglottis
*T*2   Extension to the glottis, with normal or impaired mobility
*T*3   Tumour limited to the larynx, with vocal fold fixation
*T*4   Extension through laryngeal cartilage, or into other surrounding structures

---

T2 glottic tumours which only affect cord mobility may not be detectable. This limitation is not a drawback in practice as the main point of disease staging is to exclude disease spread, i.e. to confirm that disease is not understaged clinically.

Both techniques are effective in displaying more advanced lesions. These appear as thickening of the laryngeal mucosa and on both CT and MRI contrast with submucosal fat helps outline the tumour (Figs 7.8–7.10 and 7.14) [6]. For this reason, T1-weighted sequences are preferred in MRI, although in more advanced local disease, where fat planes have been infiltrated, T2-weighted MR images may better distinguish normal and abnormal tissue (Fig. 7.9). In such cases tumour definition by CT is aided by contrast enhancement (Figs 7.12 and 7.13).

The objective of tumour staging is to provide an accurate basis for planning treatment. Localized laryngeal carcinoma may be treated by either resection or radiotherapy and practice differs between centres, some preferring resection for all early T stages, while in others total laryngectomy is reserved as a salvage procedure for patients who relapse after radiotherapy. Some surgeons routinely resect T4 lesions in view of the risk of radiation-induced chondronecrosis in infiltrated cartilage.

Studies of staging by CT and MRI in comparison with operative findings have shown both techniques to provide a good basis for treatment [12]. However MRI, by virtue of its greater soft tissue contrast and ease of displaying lesions in coronal and sagittal planes, appears to be the preferable technique when available (Figs 7.8 and 7.9) [1,13].

Disease staging has been given new impetus by the development of voice-sparing resection [6]. Tumours limited to one vocal cord may be amenable to vertical hemilaryngectomy, retaining one vocal fold or part of it. Tumours localized to the supraglottic region may be resectable by supraglottic laryngectomy, where the larynx craniad to the ventricle is removed. The precise clinical criteria for these approaches are beyond the scope of this text, but in both instances careful evaluation of the extent of disease is required, in the light of the operation being contemplated (Fig. 7.11).

In one respect MRI is clearly superior to CT. As indicated above, invasion of cartilage indicates T4 disease and its detection has a powerful effect on management. On CT non-ossified cartilage has attenuation characteristics similar to soft tissue and small degrees of invasion cannot be detected [14]. Even ossified cartilage does not provide a good basis for evaluation as the ossification pattern is usually variable. Cartilage invasion may only be apparent on CT when tumour extends through the cartilage and into adjacent tissues (Figs 7.12, 7.15 and 7.17) [6]. On MRI, however, marrow-bearing cartilage has a high signal intensity on T1-weighted sequences and is easily distinguishable from the intermediate intensity of neoplasm (Fig. 7.13). When there is little marrow in the cartilage distinction can be made on T2-weighted sequences, where tumour has a high signal intensity and the marrow cavity and cortical cartilage a low intensity [6].

An important pitfall, especially in CT, is overstaging of lesions due to accompanying inflammatory oedema, which may not be distinguishable from tumour [12]. It is helpful to be advised if laryngoscopy has shown oedema and it is important that imaging precedes laryngoscopic biopsy, or at least a week is allowed to elapse after this. An additional shortcoming of both CT and MRI is that fine structures such as the vocal folds may be stretched over the surface of adjacent masses and may be indistinguishable from the mass, suggesting they are infiltrated.

Staging examinations, whether by CT or MRI, should include the regional lymph node drainage. The glottis is poorly supplied with lymphatics and tumours tend to be well-differentiated, so that lymphadenopathy is unlikely in T1 lesions. One in four patients with T2 tumours have lymph node deposits and this rises to one in two for T3 lesions [1]. Primary supraglottic and subglottic lesions are much more likely to metastasize to lymph nodes (Fig. 7.16) [1].

The detection of lymph node spread has a powerful effect on the management of patients. However neither CT nor MRI is able to detect deposits in nodes of normal size; on both techniques the criterion of abnormality is lymph node enlargement [15]. Further, neoplastic enlargement cannot be distinguished from reactive hyperplasia due to concomitant inflammatory disease, a not uncommon situation in the upper airway.

On CT the detection of lymph node enlargement requires enhancement, which allows enlarged lymph nodes to be distinguished from adjacent vessels (Fig. 7.2). Metastases from squamous carcinoma have varying appearances, from poorly enhancing soft tissue to low attenuation material which suggests necrosis; the latter is often accompanied by a peripheral zone of enhancement (Fig. 7.18). On spin-echo MRI sequences flowing blood produces a signal void and this facilitates the detection of enlarged nodes (Fig. 7.19). Although pathologically enlarged nodes usually have a high signal intensity on T2-weighted sequences, no signal patterns specifically characterizing malignant disease have been observed.

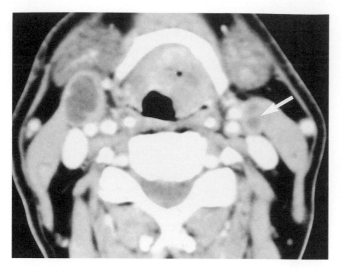

Fig. 7.18 Supraglottic carcinoma with bilateral lymphadenopathy. On enhanced CT tumour is seen to fill the pre-epiglottic space. There is an obvious lymph node metastasis on the right, with smaller metastases on the left (arrow). The nodes have an enhancing margin and a low attenuation centre, a characteristic sign in metastatic squamous carcinoma.

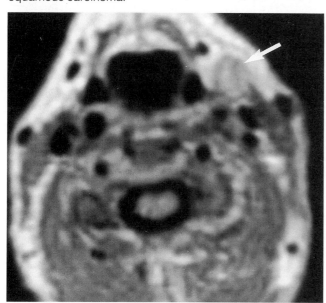

Fig. 7.19 Axial T1-weighted MRI image showing a pathologically enlarged lymph node (arrow) lateral to the left pyriform fossa. Note that on the spin echo sequence the lymph node is readily distinguished from the adjacent vessels because of the signal void created by flowing blood.

Fig. 7.20 Wegener's granulomatosis of the cricoid cartilage. (**a**) A lateral radiograph shows narrowing of the airway and increased prevertebral soft tissue, suggesting involvement of the larynx from outside. (**b**) Axial CT section through the cricoid ring shows expansion of the marrow cavity and of the submucosal tissue.

Other malignant tumours of the larynx

Other neoplasms of the larynx are rare [1]. The connective tissue elements of the larynx may give rise to sarcoma and of these, chondrosarcoma appears to be the most common, principally affecting the cricoid cartilage [10]. The laryngeal cartilages contain marrow and may be the site of neoplasia of lymphoid or myeloid origin, or of pseudotumours arising in these tissues (Fig. 7.20). Lymphoma, usually of the non-Hodgkin type, may arise in the cartilages or from any other structure of the larynx or pharynx (Fig. 7.21).

Treatment options are usually resection, radiotherapy or systemic treatment. The extent of imaging is determined by the objectives of treatment, which usually fall into two categories: either preoperative staging or a baseline for local or systemic therapy.

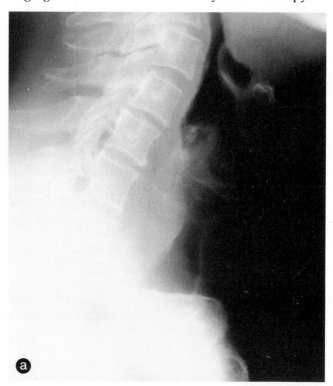

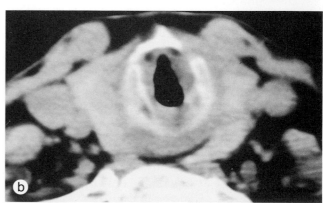

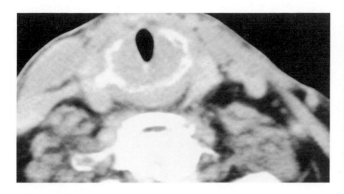

Fig. 7.21 Non-Hodgkin's lymphoma. CT section showing expansion and destruction of the cortex of the cricoid cartilage, due to non-Hodgkin's lymphoma arising within the cartilage.

Squamous carcinoma of the hypopharynx

Carcinoma of the hypopharynx carries a worse prognosis than that of laryngeal carcinoma. It is also seen in the late middle-aged and elderly and there is a preponderance in women, especially those suffering from Plummer–Vinson syndrome. Alcohol intake and cigarette smoking are also believed to be predisposing factors [10]. Generally this neoplasm presents late as the symptoms – usually persistent sore throat or pain on eating – are common to benign processes also.

This tumour, like carcinoma of the larynx, is classified according to its site of origin [16], although at presentation disease may be so advanced that this may no longer be clear.

1. The pyriform sinus, bounded laterally by the thyroid cartilage and medially by the aryepiglottic fold, is the most common site, representing approximately 60% of tumours (Figs 7.22–7.24). Carcinoma here is usually better differentiated than elsewhere in the hypopharynx.

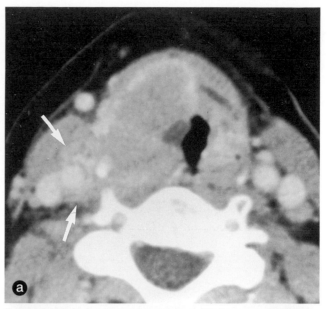

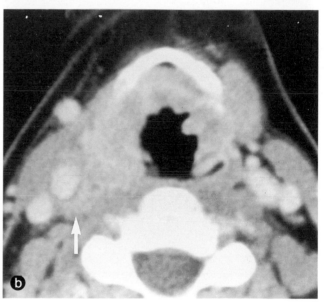

Fig. 7.23 *T*4 carcinoma of the right pyriform fossa. CT shows that this extensive tumour of pyriform fossa has spread anteriorly through the pre-epiglottic space to reach the left side. Note also the cuff of tumour surrounding the vessels on the right side (arrows). Carcinoma of the hypopharynx is frequently this extensive at presentation.

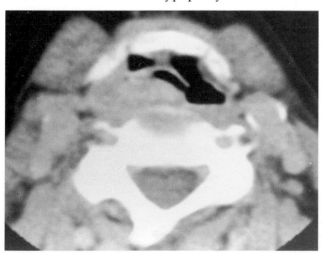

Fig. 7.22 *T*2 carcinoma of right pyriform fossa. CT shows a localized mass arising from the pyriform fossa and involving the posterior wall. The epiglottis abuts the tumour but was not involved.

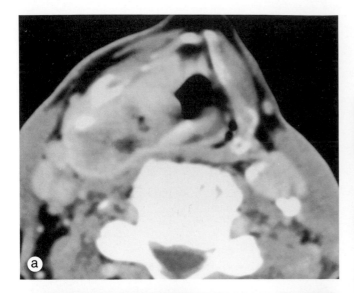

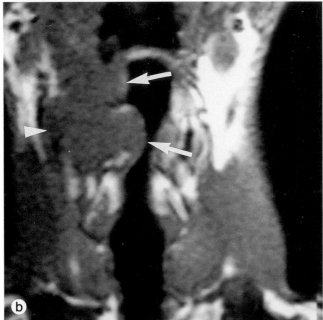

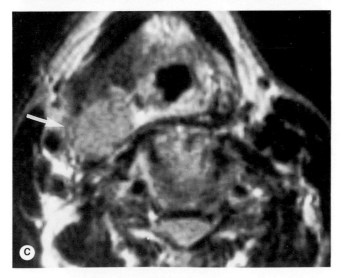

2. The postcricoid area, from the arytenoid cartilages to the lower border of the cricoid cartilage (Fig. 7.25). Tumours arising here usually carry the worst prognosis.

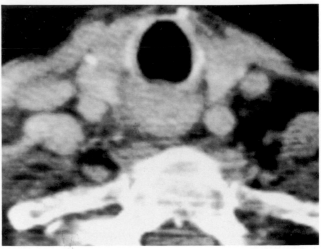

Fig. 7.25   Postcricoid carcinoma. CT shows expansion of the cricopharyngeal sphincter.

Fig. 7.24   *T*4 carcinoma; presumed origin right pyriform fossa. (a) CT scan showing a large mass occupying the right pyriform fossa and extending into the pre-epiglottic space. **(b)** Coronal MRI shows that the tumour also involves the larynx, the morphology suggesting infiltration of the true and false vocal folds and aryepiglottic fold (arrows); it is not possible to say accurately if the tumour has arisen in the larynx or the pharynx. Note that there is soft tissue abutting the vessels laterally (arrowhead). **(c)** Proton-density-weighted axial MRI image showing an inhomogeneous signal pattern within the tumour. Note that the extension to the cervical vessels (arrow) is more clearly shown than on the corresponding CT section, also that the thyroid cartilage is more readily identified on CT.

3. The posterior pharyngeal wall, from the valleculae to the level of the arytenoid cartilages (Figs 7.26 and 7.27).

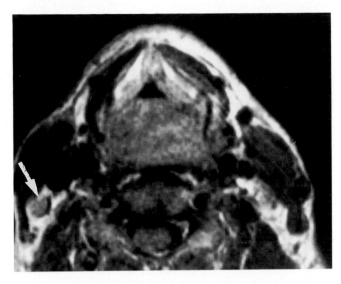

Fig. 7.26 Carcinoma of the posterior pharyngeal wall. T1-weighted axial MRI section shows a mass arising from the posterior wall of the pharynx, compressing the laryngeal vestibule. Note that there is an enlarged lymph node on the right side (arrow).

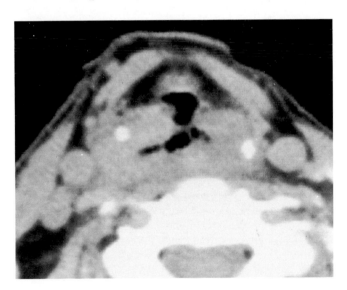

Fig. 7.27 *T4* carcinoma of the posterior pharyngeal wall. CT shows that the tumour has spread from the posterior wall into the aryepiglottic fold on each side. There is extension of tumour beyond the pharynx, into the perivascular fat bilaterally.

As in the larynx, imaging is required for disease staging rather than diagnosis, which can be made by endoscopy and biopsy. *T*-staging schemes (Table 7.2) include the anatomical classification.

Table 7.2 *T* staging scheme for carcinoma of the hypopharynx (modified from Reference 21)

T1	Tumour limited to one subsite
T2	Involvement of more than one subsite, without hemilarynx fixation
T3	Involvement of more than one subsite, with hemilarynx fixation
T4	Extension into other surrounding structures, including laryngeal cartilage

These tumours grow from mucosal plaques to invasive masses. MRI has significant advantage over CT in the detection of early disease because localized disease is of similar attenuation to the surrounding wall. Contrast enhancement has been recommended for detecting small tumours by CT but there is no doubt that MRI is preferable, on account of the high signal intensity of tumours on T2-weighted sequences. This also applies to more advanced lesions, in view of the ease on MRI of distinguishing neoplasm from the surrounding strap muscles or the thyroid gland (Fig. 7.24) [1].

Hypopharyngeal tumours are more invasive than laryngeal ones and frequently large at presentation [10]. Spread may occur medially into the larynx, laterally into the parapharyngeal tissues, superiorly to the tongue or nasopharyngeal wall or inferiorly to the oesophagus. In advanced lesions it may not be possible to say whether the tumour has arisen in the pharynx or larynx (Fig. 7.24). If so, the prognosis cannot be accurately predicted. Invasion of cartilage is common; evaluation of this follows the same rules as those for laryngeal carcinoma. Postcricoid tumours are often small but locally invasive, and all that may be detected is prominence of the sphincter (Fig. 7.25).

In most centres the invasive nature of this tumour and the probable stage at presentation are held to preclude radical surgery, although total pharyngo-laryngectomy is practised by some for early stages of disease. In most cases treatment is by radiotherapy, with salvage surgery in reserve. The staging criteria are less stringent in these circumstances, as the radiotherapist needs simply to determine the maximum extent of the lesion in order to plan therapy portals.

An important pitfall is the fact that mucosal skip lesions occur and may not be identified by imaging [1]. Effective staging therefore combines imaging with endoscopy to assess the mucosa. Apart from this, confusion may be caused by secretions pooling in the valleculae or pyriform fossa. In CT the patient can swallow between exposures, whereas on MRI avoidance of movement artefact is preferable and pooling secretions may be distressing to the patient.

Lymph node metastasis is likely in this disease and should be assumed in the majority of patients unless investigations show otherwise. It can also cover a wide area; tumours may spread laterally to the jugular nodes (Fig. 7.26), anteriorly to the prelaryngeal nodes, or superiorly in the retropharyngeal group. Postcricoid carcinoma tends to spread inferiorly to the mediastinum and examination should be extended to include this area.

Involvement by tumours of surrounding tissues

Both the larynx and pharynx may be secondarily involved by malignant tumours of other local tissues, such as thyroid carcinoma or cervical lymphadenopathy due to spread from primary tumours elsewhere, or lymphoma. In these cases the diagnosis is usually evident and investigation is required to evaluate the effect on the airway or on swallowing, or as a baseline for treatment. In children rhabdomyosarcoma is a rare but aggressive neoplasm.

Lesions which may present for diagnosis as a mass in the neck – benign or malignant – include thyroid tumours or cysts, carotid body chemodactomas, branchial cysts (Fig. 1.5, p. 8), lymphadenopathy (Fig. 7.2), or neurogenic tumours (Fig. 1.3, p. 5); these do not usually present with pharyngolaryngeal problems and are beyond the scope of this chapter. A comprehensive account will be found in Newton, Hasso and Dillon.

Investigation may also be required in patients who present with an isolated vocal fold palsy, raising the possibility of a tumour involving the recurrent laryngeal nerve (Fig. 7.28) or, less commonly, the superior laryngeal nerve. Sections must be taken through the entire course of the innervation to the larynx, which means to the level of the subclavian artery in the case of right sided palsy, or to the aortic arch if the palsy is on the left side [17].

Imaging in trauma

In the USA CT has become established in the evaluation of acute trauma to the larynx when this results from direct blows, for example from steering wheel injuries [18]. The main objective is to detect fractures of the laryngeal cartilages and collapse of the airway. Obstruction due to haematoma can be distinguished from more serious damage, and arytenoid cartilage dislocation (Fig. 7.29) may also be diagnosed. In most of these instances acute investigation of airway damage, which requires careful patient monitoring in the gantry, is justified only if the patient is being considered for urgent airway repair.

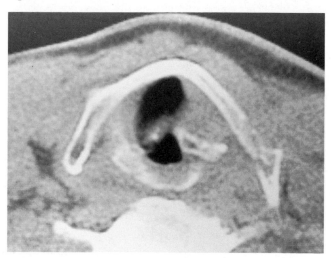

Fig. 7.29 Laryngeal trauma. In this patient the larynx was damaged by hyperextension of the neck during a road traffic accident. The left arytenoid cartilage is dislocated anteriorly and the cord paralysed in partial adduction. Fracture of the left thyroid lamina has also occurred.

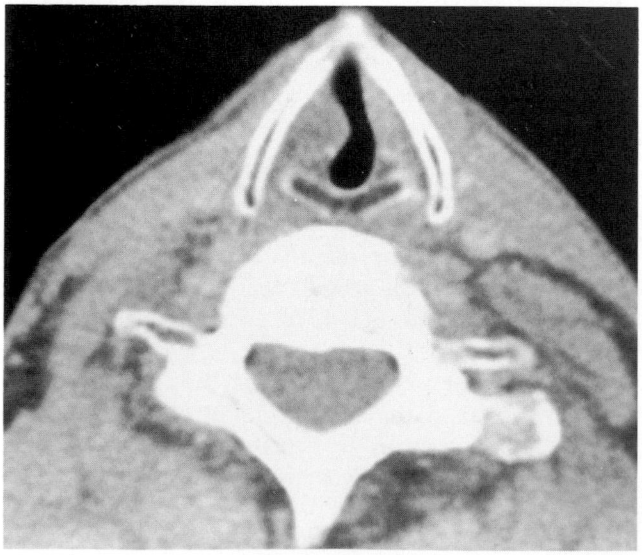

Fig. 7.28 Right vocal fold paralysis in a patient with recurrent laryngeal nerve infiltration due to carcinoma of the thyroid. In this condition the arytenoid cartilage is displaced anteriorly, so that the cord lies partly adducted. Note that the attenuation value of the cord is reduced due to atrophy of the intrinsic muscles.

Where laryngeal injuries are managed conservatively by tracheostomy but delayed reconstruction is considered, imaging by either CT or MRI is a good basis for planning the surgical approach. In particular, the important distinction between an airway which has collapsed due to post-traumatic chondronecrosis (Fig. 7.30) and one which is constricted by surrounding scarring can be made with confidence.

Laryngoceles usually present for diagnosis of a mass in the neck, or voice change, and are classified as internal or external, according to their extent. Both types are due to outpouching of the mucosa of the laryngeal ventricle. An internal laryngocele passes laterally and superiorly in the submucosa, being contained within the laryngeal skeleton. The external variety pass out of the larynx through the neurovascular space in the thyrohyoid membrane, to enter the soft tissues of the neck (Fig. 7.31). Either

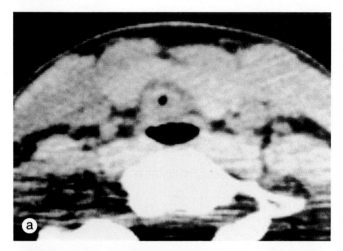

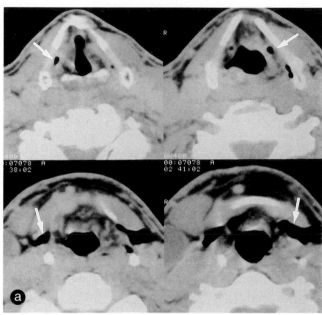

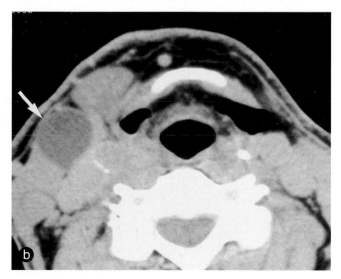

Fig. 7.30 Evaluation prior to airway reconstruction. This patient has collapse of the airway due to chondronecrosis occurring after a steering-wheel injury to the larynx. (**a**) Axial CT shows that the tracheal wall has collapsed, with minimal fibrosis; a situation which is a good basis for surgery. (**b**) Coronal CT reconstruction showing the longitudinal extent of the stricture.

Congenital lesions and benign lesions

A number of congenital malformations of the larynx exist but are likely to be diagnosed by endoscopy. This is also true of laryngeal cysts in children, which may be related to laryngoceles in the adult [1].

Fig. 7.31 CT of laryngocele. (**a**) Four sections from the false vocal folds to the hyoid, showing the typical track of bilateral laryngoceles, arising from the ventricle and piercing the false fold to run superiorly in the pre-epiglottic space, pierce the thyrohyoid membrane and emerge in the subcutaneous fat at the neck posterior to the strap muscles (arrows). (**b**) In the same patient the distal extent of the right laryngocele was cystic (arrow).

type may contain air, or be full of secretions [6].

Laryngoceles may be diagnosed on both CT and MRI the air-filled type are easier to assess on CT (Fig. 7.31) [19]. The diagnosis is confirmed by finding an air-containing space which can be traced inferiorly to the level of the laryngeal ventricle. The origin of cystic lesions may be more difficult to define if the track is collapsed and if only the external element is detected; these may be confused with other benign cysts of the neck.

A wide range of benign or congenital cysts occurs in the tissues of the neck adjacent to the larynx and hypopharynx. These almost all appear as simple thin-walled cysts on imaging studies and are distinguished by their anatomical site. Of the more common, cysts of the second branchial cleft are usually near the submandibular gland. Cystic hygromas usually occur lower in the neck, in the supraclavicular fossa. Thyroglossal cysts may arise anywhere between the tongue and the thyroid gland, and may therefore be associated with the anterior aspect of the larynx; rarely they occur in an anterior defect in the thyroid cartilage.

Imaging following therapy

Significant changes in appearances take place in the larynx after radiotherapy. The inflammatory response to treatment produces oedema throughout the treatment field and this obscures fat planes and the primary tumour, making further evaluation difficult (Fig. 7.32) [1,13]. This reaction usually settles within a few months and any area of persistence should be regarded as suspicious of residual neoplasm. Radiotherapy also reduces the cellular content of marrow within the cartilages, with the result that after the inflammatory response has resolved the appearance of the marrow approaches that of normal fat (Fig. 7.33).

Imaging after radical resection is usually directed at possible recurrence. The anatomy can be altered dramatically after resection (Figs 7.34 and 7.35) and a knowledge of the surgical details is essential to accurate investigation (Fig. 7.36) [20]. In particular, accurate assessment of the findings after total laryngectomy and pharyngectomy needs to take into account whether pharyngeal reconstruction has taken place or not. Recurrent masses may be revealed in areas of fibrosis by the fact that they enhance on CT, or by T2-weighted MRI sequences which reveal recurrence as an area of high intensity. In both techniques false positive examinations can be the result of continuing inflammation, whether due to infection or in the first 6 months after radiotherapy [1].

When post-treatment inflammation has settled MRI is valuable for distinguishing prominent scars from recurrent neoplasm because these have different signal characteristics, inactive scars being of low signal intensity on all MRI sequences.

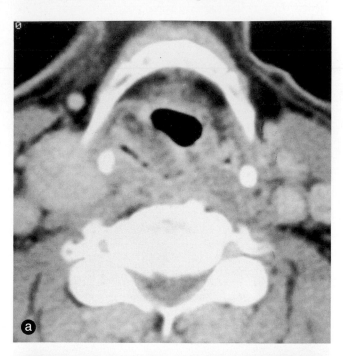

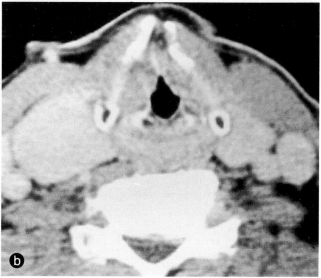

Fig. 7.32 The effect of radiotherapy in a patient undergoing treatment. (**a**) CT section through the aryepiglottic fold, showing oedema of the fold and the pre-epiglottic space. (**b**) Section through the false vocal folds showing that these are rendered featureless by inflammatory oedema.

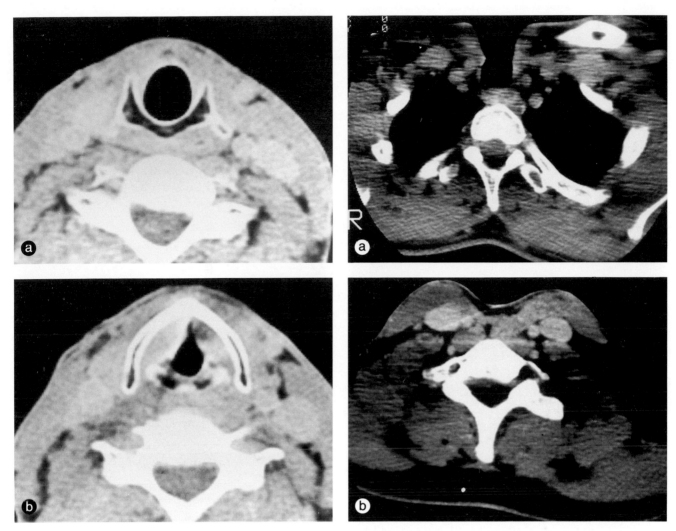

Fig. 7.33   The delayed effects of radiotherapy. This patient underwent radiotherapy for squamous carcinoma of the right vocal fold one year previously. Sections through the cricoid cartilage (**a**) and the thyroid cartilage (**b**) show decreased attenuation of the medullary cavity, due to loss of cellular marrow. Note that there is increased attenuation in the right vocal fold, due to recurrent neoplasm.

Fig. 7.34   Appearances in tracheostomy and total pharyngectomy. (**a**) CT section at the level of the stoma. (**b**) CT section at the level of the thyroid gland, showing only a small fat space at the site of the pharynx, with medial displacement of the thyroid lobes into the surgical defect.

Fig. 7.35   CT section showing the appearances following total pharyngectomy and intestinal transplant. In this patient the anatomy of the thyroid gland is preserved and the transplant is surrounded by fat and accompanying blood vessels.

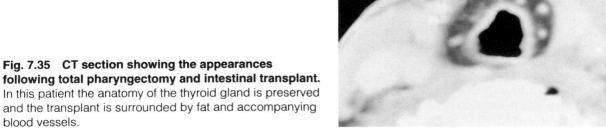

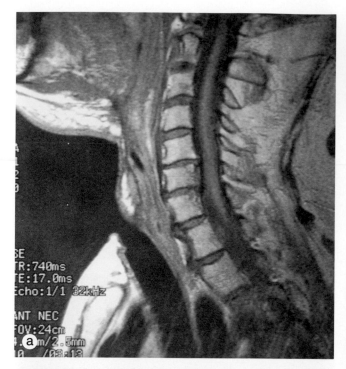

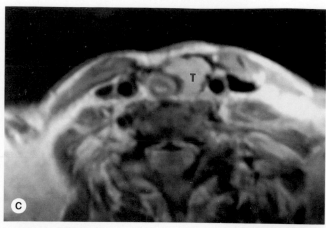

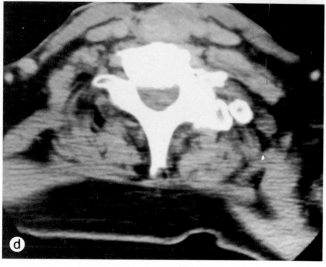

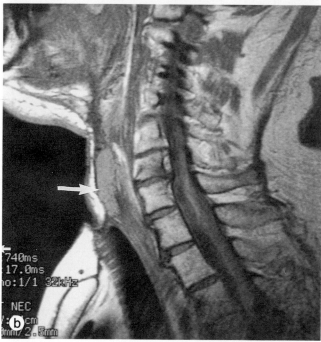

Fig. 7.36 Appearances following tracheostomy and total laryngectomy, with preservation of the pharynx.
(**a**) Midline T1-weighted sagittal MRI section showing the tracheostomy. (**b**) A sagittal MRI section to the left of the midline shows an apparent soft tissue mass (arrow).
(**c**) Axial proton-density-weighted MRI section shows that the right lobe of the thyroid gland has been resected but that the mass seen on sagittal sections was due to residual left thyroid gland (T). (**d**) Axial CT section for comparison with (**c**), showing that MRI better distinguishes the pharynx and residual thyroid gland.

ACKNOWLEDGMENTS

I am grateful to my colleagues Mrs S. Ainslie, Mrs
W. Jackson, Mrs C. Westbrooke, Mr S. Dezonie and
Dr N. R. Moore for provision of the images of
normal subjects, and to the staff of the Oxford MRI
Centre and Regional CT unit for their support in this
work; also to Mrs B. Walters for secretarial help.

REFERENCES

1. Hanafee WN, Ward PH The larynx. In: Hanafee W, Ward P, eds. Clinical correlations in the head and neck, vol. 1. New York: Thieme Medical Publishers, 1990.
2. Jones B, Donner MW. Normal and abnormal swallowing: imaging in diagnosis and therapy. New York: Springer Verlag, 1991.
3. Noyek AM, Shulman HS, Steinhardt MI. Contemporary laryngeal radiology – a clinical perspective. J Otolaryngol 1982; 11: 178–190.
4. Semenkovich JW, Balfe DM, Weyman PJ et al. Barium pharyngography: comparison of single and double contrast. AJR 1985; 144: 715–720.
5. Gritzmann N, Traxler M, Grasl M, Pavelka R. Advanced laryngeal cancer: sonographic assessment. Radiology 1989; 171: 171–175.
6. Curtin HD. Imaging of the larynx: current concepts. Radiology 1989; 173: 1–11.
7. Jeffrey RB, Dillon WP. The larynx. In: Newton TH, Hasso AN, Dillon WP, eds. Computed tomography of the head and neck. New York: Raven Press, 1988.
8. Silverman PM, Johnson GA, Korobkin M. High-resolution sagittal and coronal reformatted CT images of the larynx. AJR 1983; 140: 819–822.
9. Towler CR, Young SW. Magnetic resonance imaging of the larynx. Magnetic Resonance Q 1989; 5: 228–241.
10. Batsakis J. Tumours of the head and neck: clinical and pathological considerations. Baltimore: Williams and Wilkins, 1979.
11. Reid MH. Laryngeal carcinoma: high-resolution computed tomography and thick anatomic sections. Radiology 1984; 151: 689–696.
12. Silverman PM, Bossen EH, Fisher SR et al. Carcinoma of the larynx and hypopharynx: computed tomographic-histopathologic correlations. Radiology 1984; 151: 697–702.
13. Lufkin RB, Hanafee WN, Wortham D, Hoover L. Larynx and hypopharynx: MR imaging with surface coils. Radiology 1986; 158: 747–754.
14. Mafee MF, Schild J, Michael AS et al. Cartilage involvement in laryngeal carcinoma: correlation of CT and pathologic macrosection studies. J Comput Assist Tomography 1984; 8: 969–973.
15. Som P. Lymph nodes of the neck. Radiology 1987; 165: 593–600.
16. Dillon WP, Miller EM. Cervical soft tissues. In: Newton TH, Hasso AN, Dillon WP, eds. Computed tomography of the head and neck. New York: Raven Press, 1988.
17. Glazer HS, Aronberg DJ, Lee JKT, Sagel SS. Extralaryngeal causes of vocal cord paralysis: CT evaluation. AJR 1983; 141: 527–531.
18. Jeffrey RB. CT of laryngeal trauma. In: Federle MP, Brant-Zawadzki M, eds. Computed tomography in the evaluation of trauma. Baltimore: Williams and Wilkins, 1986.
19. Glazer HS, Mauro MA, Aronberg DJ et al. Computed tomography of laryngoceles. AJR 1983; 140: 549–552.
20. DiSantis DJ, Balfe DM, Hayden RE et al. The neck after total laryngectomy: CT study. Radiology 1984; 153: 713–717.
21. Spiessl B, Beahrs OH, Hermanek P et al. Union Internationale contre le cancer TNM Atlas: illustrated guide to the TNM/pTNM classification of malignant tumours. Berlin: Springer Verlag, 1990.

FURTHER READING

Anzai Y, Lufkin R, Jabour B, Hanafee W. The larynx. In: Higgins CB, Hricak H and Helms CA, eds. Magnetic resonance imaging of the body. 2nd ed. New York: Raven Press, 1992.
Newton TH, Hasso AN, Dillon PD. Computed tomography of the head and neck. New York: Raven Press, 1988.

Index

Page numbers printed in *italic* represent tables, those printed in **bold** represent figures.